The U.S. Environmental Protection Agency (EPA) is pleased to issue the Healthy Buildings, Healthy People (HBHP) report, a vision for indoor environmental quality in the 21st Century. The importance of the indoor environment to human health has been highlighted in numerous environmental risk reports, including the 1997 report of the Presidential and Congressional Commission on Risk Assessment and Risk Management. On average, we spend about 90 percent of our time indoors, where pollutant levels are often higher than those outside. Indoor pollution is estimated to cause thousands of cancer deaths and hundreds of thousands of respiratory health problems each year. In addition, hundreds of thousands of children have experienced elevated blood lead levels resulting from their exposure to indoor pollutants.

The report challenges all of us to work together to improve the quality of our environment. It can also serve as the basis for discussion and education among professionals in public policy, health, building sciences, product manufacturing, and environmental research. The report is also a blueprint for channeling available resources. Already, EPA has undertaken program initiatives focusing on childhood asthma, characterizing the effect of building and consumer products on the indoor environment, increasing the demand for cleaner indoor products for use in schools, creating standards of care for existing buildings, and designing guidance for new schools. EPA is also integrating good indoor environmental quality (IEQ) concepts into the Energy Star® label program for commercial office buildings. Moreover, other current federal programs, while not direct outgrowths of the HBHP effort, are complementary of it. For example, at the Department of Housing and Urban Development (HUD), the Healthy Homes project has identified moisture and molds as a priority to be addressed in its grants process, and the Healthy People 2010 effort at the Department of Health and Human Services (DHHS) contains several goals relating to IEQ progress in existing buildings. We challenge others, including government, tribes, the health community, academia, non-profit organizations, and industry, to embrace the HBHP goals and work together to invest in the actions outlined in the report. In this way, we can begin to make progress towards realizing the vision of HBHP.

The HBHP report is the outcome of a cross-Agency effort to define a strategic vision and potential actions for improving the quality of our indoor environment and was jointly led by the Office of Air and Radiation (OAR) and the Office of Prevention, Pesticides, and Toxic Substances (OPPTS) with substantial involvement from the Office of Research and Development (ORD). As part of this effort, we sought the advice of many outside experts and visionaries.

During this collaborative process, we learned a great deal from our stakeholders. For example, we need to further understand indoor sources of pollutants and their health effects, integrate building design and maintenance, encourage federal buildings to be "model" indoor environments, support the development of new product technologies, and educate the public. Also, we need to work closely with other federal agencies; state, local and tribal governments; health and community organizations; and industry and other private groups to improve the Nation's health.

Based on stakeholder and cross-Agency input, EPA issued a draft report in March 2000 containing a vision, goals, guiding principles, and potential actions to improve human health indoors. The draft report was available on EPA's web site and was sent to over 300 stakeholders representing the public, environmental and health interests, academia, federal agencies, state and local governments, tribes, non-profit organizations, trade associations, and industry. The public was asked to submit comments by May 31, 2000. This comment period was extended to June 30, 2000 at the request of several commentors. We received comments from over 40 individuals and organizations; many of them have been incorporated into the final report, or have been addressed in Appendix C. Interestingly, this effort has attracted attention in Canada where a parallel effort, Healthy Indoors: Achieving Healthy Indoor Environments in Canada (www.HealthyIndoors.com) is using our draft report as a centerpiece in its stakeholder dialogues.

The draft HBHP report was positively received, and many indicated that the document was a significant step in addressing an important, but often overlooked, public health issue. Although there were a number of specific suggestions for changes to the draft report, nearly all commentors felt the report was comprehensive, and that the vision and goals captured the central themes and needs of the issue. Several indicated that the potential actions contained in the draft report were strategic, and that, when implemented, would be helpful in addressing the quality of our indoor environment. A summary of the comments is contained in Appendix C.

While many of the comments we received were incorporated, the basic structure of the draft report has been maintained in the final HBHP report. Chapter 1 focuses on why human health indoors deserves the scrutiny, concern, and action of policy makers. These reasons are primarily health-related. Health risks associated with indoor environments include asthma, cancer, and reproductive and developmental effects. However, significant gaps still exist in the current state of knowledge about indoor environmental risks and exposures. We also believe that a particular emphasis must be placed on children's health. Chapter 2 presents a vision statement and outlines goals, broad strategies, and guiding principles to achieve success in every sector of our society over the next 25 to 50 years. In short, our objective is to realize major human health gains over the next 50 years by upgrading indoor environments. Five goals or strategies have been set to accomplish this objective: (1) achieve major health gains and improve professional education; (2) foster a revolution in the design of new and renovated buildings; (3) stimulate nationwide action to enhance health in existing structures; (4) create and use innovative products, materials, and technologies; and (5) promote health-conscious individual behavior and consumer awareness. In addition to providing information on actions and strategies that can be taken to protect people indoors, EPA's vision acknowledges the important role played by individuals in protecting their own health and the health of those around them. Chapter 3 lays out potential actions that EPA or others may pursue.

Appendix A provides an overview of current indoor environmental program priorities in various offices within EPA. Appendix B examines the roles of the Agency's partners in indoor environmental protection, including federal, state, local, and tribal organizations, as well as stakeholders in the private sector. Appendix C provides a summary of the comments on the draft report and how the comments can be accessed through our docket.

ACGIH	American Conference of Governmental Industrial Hygienists
AIA	American Institute of Architects
AIDS	Acquired Immune Deficiency Syndrome
AMCL	Alternative Maximum Contaminant Level
ANSI	American National Standards Institute
ASHRAE	American Society of Heating, Refrigerating, and Air-Conditioning Engineers
ASTHO	Association of State and Territorial Health Officials
ASTM	American Society for Testing and Materials
ATSDR	Agency for Toxic Substances and Disease Registry
BEIR	National Academy of Sciences' Committee on the Biological Effects of Ionizing Radiation
CDC	Centers for Disease Control and Prevention
CDN	Clinical Directors Network
CDPHE	Colorado Department of Public Health and the Environment
CERCLA	Comprehensive Environmental Restoration Compensation and Liability Act
CLI	Consumer Labeling Initiative
CO	Carbon Monoxide
CPSC	U.S. Consumer Product Safety Commission
DHHS	U.S. Department of Health and Human Services
DINP	Diisononyl Phthalate
DOC	U.S. Department of Commerce
DOE	U.S. Department of Energy
DOL	U.S. Department of Labor
ECOS	Environmental Council of the States
EPA	U.S. Environmental Protection Agency
ETS	Environmental Tobacco Smoke
ETV	Environmental Technology Verification
FCIC	Federal Consumer Information Center
FHA	Federal Housing Administration
FIFRA	Federal Insecticide, Fungicide, and Rodenticide Act
GSA	General Services Administration
HBHP	Healthy Buildings, Healthy People
HMOs	Health Maintenance Organizations
HUD	U.S. Department of Housing and Urban Development
HVAC	Heating, Ventilating, and Air-Conditioning
IAQ	Indoor Air Quality
IEQ	Indoor Environmental Quality
IQ	Intelligence Quotient
IUR	TSCA Inventory Update Rule
MCL	Maximum Contaminant Level

MHE	Master Home Environmentalist
NAHB	National Association of Home Builders
NAHB-RC	National Association of Home Builders-Research Center
NAS	National Academy of Sciences
NCEA	EPA National Center for Environmental Assessment
NCI	National Cancer Institute
NCSL	National Conference of State Legislatures
NEHA	National Environmental Health Association
NERL	EPA National Exposure Research Laboratory
NGA	National Governors Association
NH COSH	New Hampshire Coalition for Occupational Safety and Health
NHEERL	EPA National Health and Environmental Effects Research Laboratory
NIH	National Institutes of Health
NIOSH	National Institute for Occupational Safety and Health
NIST	National Institute of Standards and Technology
NRMRL	EPA National Risk Management Research Laboratory
OAR	EPA Office of Air and Radiation
OARM	EPA Office of Administration and Resources Management
OCHP	EPA Office of Children's Health Protection
OECA	EPA Office of Enforcement and Compliance Assurance
OEJ	EPA Office of Environmental Justice
OGWDW	EPA Office of Ground Water and Drinking Water
OPEI	EPA Office of Policy, Economics, and Innovation
OPPTS	EPA Office of Prevention, Pesticides, and Toxic Substances
ORD	EPA Office of Research and Development
OSHA	Occupational Safety and Health Administration
OSHAct	Occupational Safety and Health Act
OSW	EPA Office of Solid Waste
OSWER	EPA Office of Solid Waste and Emergency Response
OW	EPA Office of Water
PBT	Persistent, Bioaccumulative, and Toxic
PESP	Pesticide Environmental Stewardship Program
PHSA	Public Health Services Act
PTI	Public Technology, Inc.
SBS	Sick Building Syndrome
SIDS	Sudden Infant Death Syndrome
TSCA	Toxic Substances Control Act
UL	Underwriters Laboratories
USDA	U.S. Department of Agriculture
VA	Veteran's Administration
VOC	Volatile Organic Compound

CONTENTS

REFERENCES

APPENDIX A

APPENDIX B

APPENDIX C

Chapter 1

WHY STUDY HUMAN HEALTH INDOORS?

BACKGROUND

Americans spend about 90 percent of their time indoors, where concentrations of pollutants are often much higher than those outside. Risk assessments performed for radon, environmental tobacco smoke (ETS), and lead have shown that health risks are substantial. Thousands of chemicals and biological pollutants are found indoors, many of which are known to have significant health impacts both indoors and in other environments. Although much is known or suspected regarding human health risks in the indoor environment, a comprehensive, integrated effort to assess and manage indoor risks has yet to be undertaken.

In 1987, the EPA Comparative Risk Project was conducted to examine the relative risk of environmental problems. In 1990, the Relative Risk Reduction Strategies Committee of EPA's Science Advisory Board conducted a similar, extensive analysis of relative environmental risk. Both resulting reports, *Unfinished Business: A Comparative Assessment of Environmental Problems* (U.S. EPA 1987) and *Reducing Risk: Setting Priorities and Strategies for Environmental Protection* (U.S. EPA 1990), ranked indoor air pollution among the top five environmental risks to public health. In 1997, the Presidential and Congressional Commission on Risk Assessment and Risk Management also found that indoor environmental pollution can pose a substantial environmental risk and advised EPA to address those risks. During the release of its report, the Commission chairman highlighted indoor environmental pollution as one of the greatest risks to human health.

Americans are concerned about their own health and the health of their children. However, despite efforts by EPA and other private and public groups to conduct research on indoor environmental issues and to communicate the findings of that research, most Americans do not have a clear sense of the significant health risks of indoor pollution. They also do not know what they can do to reduce risk for asthma, cancer, and other serious diseases caused by indoor pollutant exposure.

Nor do many building professionals yet understand how to integrate indoor air quality objectives into the design and operation of the Nation's buildings. The economic value of improved health and productivity can be substantial, and can be achieved through integrated building design, commissioning, and operations which may reduce costs or result in only modest cost increases. Thus, indoor air quality promises to become an important part of the movement toward green buildings and green products. Further, any productivity gains will serve to enhance the Nation's competitiveness in the global economy.

PRINCIPLES FOR HBHP

The following two principles will serve to provide a workable context for identifying and addressing priorities for improving the indoor environment:

First, exposure needs to occur within or be aggravated by the building.

This principle is relatively straightforward. However, there are diverse types of buildings, including homes, schools, day care facilities, nursing homes, offices, factories, hospitals, hotels, restaurants, retail shops, theaters, arenas, and correctional facilities. Impacts on human health and methods for reducing exposure to indoor air pollution and the associated risk vary by building type, use, and activity.

Second, risk reduction must be accomplished through better building design, construction, and operation; improvements in the development and use of indoor products; or mitigation of existing exposures within a building or in its immediate vicinity.

This principle excludes some risks that, although they occur indoors, originate outside the building and are best mitigated at a distance. For example, risks would be excluded if the source of the pollutant is industrial discharge (e.g., drinking water contaminated by lead tailings from a mine or air pollutants entering the environment from industrial smokestacks). Risks would be included when the pollutant is added indoors (e.g., drinking water contaminants from lead solder in plumbing in the building or air pollutants emitted from sources within the building). Pesticide residues on food from the spraying of crops would be excluded, while pesticides used directly indoors, or that are used near the home and are tracked indoors, would be considered indoor pollution.

INDOOR HUMAN HEALTH RISKS

The risks to human health indoors include asthma, cancer, reproductive and developmental problems, and other health effects.

Exposures to radon, ETS, lead, and other chemical and biological contaminants in the indoor environment result in a wide array of health impacts. Known health effects of indoor pollutants include asthma; cancer; developmental defects and delays, including effects on vision, hearing, growth, intelligence, and learning; and effects on the cardiovascular system (heart and lungs). Pollutants found in the indoor environment may also contribute to other health effects, including those of the reproductive and immune systems. Some pollutants, such as carbon monoxide (CO), are acutely toxic and can result in death. The following sections summarize several health endpoints of greatest concern.

ASTHMA

An estimated 17 million Americans suffer from asthma (U.S. EPA 1999). In addition, about 5,000 deaths occur yearly from asthma—an increase of 33 percent in the last decade (Mannino et al. 1998). Consequently, the social and economic costs are large. Among chronic diseases, asthma is the number one cause of absenteeism from school (Pope et al. 1993). Asthma cost an estimated $6.2 billion in the United States in 1990, including direct medical and indirect non-medical costs combined (Weiss et al. 1992). An update of this figure would fall in the range of $7 to $9 billion in 1998 dollars.

Some groups in this country (e.g., children, certain minorities, seniors, and low-income, urban populations) are disproportionately affected by asthma. An estimated 1.8 million people required emergency room services for asthma in 1995. Mortality rates associated with asthma among African-Americans, as a whole, are two- to three-fold higher than those among whites. Mortality rates for African-American children are five-fold higher than those for their white peers (Mannino et al. 1998). While research has not yet explained the rise in the incidence of asthma, nor all the reasons why individuals first contract it, there is general agreement that controlling indoor exposures is an important protective measure (NAS 2000).

Recently, the National Academy of Sciences (NAS)/Institute of Medicine issued a report on asthma and indoor air quality, confirming that dust mites and other allergens, microorganisms, and some chemicals found indoors are triggers for asthma. In addition, the report stated there was sufficient evidence to link the exposure of preschool-aged children to ETS and exposure to house dust mites with the development of asthma (NAS 2000). ETS may significantly aggravate symptoms of asthma for 200,000 children and may affect as many as 1,000,000 children to some extent (U.S. EPA 1992).

Cancer

A number of indoor contaminants, such as asbestos, radon, tobacco smoke, and benzene, are known human carcinogens. Other indoor contaminants, such as certain chlorinated solvents, polycyclic aromatic hydrocarbons, aldehydes, and pesticides, are considered likely to cause cancer in humans.

The National Academy of Sciences, in its latest report on radon health science (NAS 1998), concluded that radon is the second leading cause of lung cancer in the country. NAS has estimated that about 12 percent of the lung cancer deaths in the United States are linked to radon. They calculate the number of lung cancer cases attributable to radon exposure to range from 15,000 to 22,000 annually.

Environmental tobacco smoke is estimated to cause an additional 3,000 lung cancer deaths in non-smokers each year (U.S. EPA 1992). Other forms of cancer have also been found to be associated with indoor pollutants (e.g., leukemia with benzene; bladder cancer with ETS).

Reproductive and Developmental Effects

During the period 1991-1994, almost 900,000 children aged 1-5 years had elevated blood lead levels, which are associated with a variety of developmental delays, including decreased intelligence quotient (IQ); stature, growth, and hearing deficits; and learning disabilities (U.S. DHHS 1997a). The geometric mean blood lead level for children aged 1-5 years was 2.7 ug/dl in 1991–1994. In 1999, the geometric mean was estimated to be 2.0 ug/dl for this age group. The 1999 sample was not large enough to produce reliable estimates of the number of children with elevated blood lead levels. State surveillance data are consistent with the decline in the national geometric mean, but the state data also confirm that the risk for an elevated blood lead level in children remains high in some counties and varies greatly among and within states (U.S. DHHS 2000). Several studies indicate that common indoor pollutants such as lead and ETS can also impair fetal development. A California report estimates that 9,700 to 18,600 cases of low birth weight in infants are caused each year by ETS (NCI 1999).

Many other environmental agents, including a number of chemicals commonly found indoors (e.g., tobacco smoke, some pesticides, lead and other heavy metals, alcohols, plastic additives), are suspected of causing developmental toxicity in humans (U.S. EPA 1991a, NCI 1999). Endocrine disruptors (e.g., certain pesticides and plasticizers), which affect the normal function of sex and thyroid hormones, present a new area of concern for reproductive toxicity. Adverse effects on a developing child may result from exposure prior to conception in either parent, exposure during pregnancy, or post-natal exposure. These effects range from low birth weight to genetic diseases to lower IQs and infertility.

OTHER HEALTH EFFECTS

Indoor environments can cause or amplify many other health effects as well. The California ETS report estimates that 35,000 to 62,000 cardiovascular deaths per year among non-smokers can be attributed to ETS exposure (NCI 1999). Recent studies have shown that, compared to those who had not been exposed, ETS was associated with a 20 percent increase in the progression of atherosclerosis (hardening of the arteries) (Howard et al. 1998). Carbon monoxide poisoning associated with the improper use and maintenance of fuel-burning appliances kills more than 200 people per year in this country and results in about 10,000 admissions to hospital emergency rooms for treatment (U.S. CPSC 1997). An additional 600 to 700 accidental deaths from CO poisoning occur indoors from other sources, including automobiles (Cobb and Etzel 1991). The agent for Legionnaires' disease, a potentially deadly pneumonia which affects 10,000 to 15,000 people each year, is associated with cooling systems, whirlpool baths, humidifiers, food market vegetable misters, and other indoor sources, including residential tap water (EPA et al. 1994; U.S. DHHS 1997b). Effects associated with toxins from indoor fungi and bacteria range from short-term irritation to immunosuppression and cancer (EPA et al. 1994).

Studies show that symptoms of sick building syndrome (SBS) may be caused or intensified by indoor environmental problems (U.S. EPA 1991b, U.S. EPA et al. 1994). The term "sick building syndrome," first employed in the 1970s, describes a spectrum of specific and non-specific complaints reported by a population of building occupants. These symptoms can be associated with their presence in the building. These complaints may also result from causes other than SBS, including illness contracted outside the building, acute sensitivity (e.g., allergies), job-related stress or dissatisfaction, and other factors. Data are insufficient to thoroughly evaluate many SBS problems.

UNCERTAINTIES

Although EPA has estimated the carcinogenic potency of a number of indoor pollutants, the Agency has conducted comprehensive population risk assessments for only a few substances (e.g., radon, ETS, lead). A comprehensive indoor environments risk assessment should cover all of the chemical and biological indoor pollutants for which sufficient toxicological and exposure data exist.

Most chemicals in commercial use have not been tested for possible health effects. Fewer than one-third of regulated, high-production chemicals, including many found indoors, have undergone even a screening level of testing for adverse effects. Health effects data are particularly critical for indoor exposure because median indoor concentrations are one to five times the median outdoor concentrations of many hazardous air pollutants. Considering that people spend approximately 90 percent of their time indoors, median indoor exposures (concentration multiplied by time) may be 10 to 50 times higher than outdoor exposures (U.S. EPA 1998).

Significant uncertainties exist in the areas of exposure assessment and control. For example, data are lacking on the rate and frequency of emissions from many sources, such as building materials and consumer products. There is also a lack of data on the identity of the chemicals emitted, as well as on the cost and performance of solutions to reduce exposures. While there are standard methods to quantify emissions from certain types of products and materials (e.g., carpets, office furniture, paints), many more are needed to facilitate widespread commercial development of new products and materials that emit significantly lower levels of indoor pollutants. Significant uncertainties still exist regarding how a change in building design, operation, and maintenance will influence the mix of indoor pollutants, as well as how to measure the concentrations of biological contaminants present indoors. Exposures in schools, residences, and most other non-occupational indoor environments still remain largely unstudied.

WHO IS MOST AT RISK?

Children often experience higher exposures to environmental pollutants than adults because, per pound of body weight, they breathe more air and ingest more material than adults. Children also more readily absorb contaminants. Additional exposure pathways resulting from activities such as crawling and sucking and gnawing on toys can also elevate risk for children. For example, between 1991 and 1994, almost 900,000 children in this country had unacceptable blood lead levels from exposure in their own homes (U.S. DHHS 1997a). Minority status, income status, and age of housing have all been shown to correlate with elevated blood lead in children. Children are more susceptible to the effects of lead exposure because their brains are still developing, they ingest more lead than adults through hand-to-mouth activity, and their developing systems more readily absorb lead than those of adults (U.S. EPA 1996).

EPA estimates that ETS is responsible for between 150,000 and 300,000 lower respiratory tract infections in infants and children under 18 months of age, as well as an increased prevalence of fluid build-up in the middle ear. This is estimated to result in between 7,500 and 15,000 hospitalizations each year. Post-natal ETS exposure has also been implicated in 1,900 to 2,700 cases of sudden infant death syndrome (SIDS) annually (NCI 1999).

Individuals may be more vulnerable to indoor contaminants because of age, genetics, nutrition, metabolism, exposure levels, existing diseases, and other factors. For example, older people are at particular risk for adverse effects on the nervous and cardiovascular systems; asthmatics are more vulnerable to allergens and respiratory irritants; and people with acquired immune deficiency syndrome (AIDS) and other immunodeficiencies are more vulnerable to pneumonia, pathogenic yeasts, and other illnesses.

END NOTES

[1]When attempting to reduce the "total" impact on human health, knowledge of the relative risk from ambient air pollutants that make their way indoors and from pollutants emitted by indoor sources will determine the focus of where the most effective risk reduction can occur.

[2]A U.S. District Court decision vacated several chapters of EPA's 1993 scientific risk assessment document that served as the basis for EPA's classification of secondhand smoke as a Group A carcinogen and estimates that ETS causes 3,000 lung cancer deaths in non-smokers each year. The ruling was largely based on procedural grounds. EPA is appealing this decision. None of the findings concerning the serious respiratory health effects of secondhand smoke in children was challenged.

EPA firmly maintains that the bulk of the scientific evidence demonstrates that ETS causes lung cancer and other significant health threats to children and adults. EPA's 1993 report estimating the risks posed by ETS was peer-reviewed by 18 eminent, independent scientists who unanimously endorsed the study's methodology and conclusions. Since then, numerous independent health studies have presented an impressive accumulating body of evidence that confirms and strengthens the EPA findings. It is widely accepted in the scientific and public health communities that secondhand smoke poses significant health risks to children and adults.

CHAPTER 2

VISION AND GOALS

VISION

STRATEGIES

- Leveraging Action Through Partnerships
- Encouraging New Technologies
- Market Incentives
- Research and Development
- Legislation, Policy, and Standards
- Education and Information

All across our nation, people live, work, and learn in healthy indoor environments. The environments inside our buildings help us reach our full potential for good health and productivity. No one is excluded: we create healthy buildings at every income level and help all our children grow up to be healthy adults. We understand the importance of healthy indoor environments, create a demand for them, and expect them as something that everyone deserves. By choosing designs, ventilation systems, materials, and products wisely, we are able to create healthy buildings while substantially reducing energy use, cutting materials costs, and raising productivity. The Nation's success in improving human health indoors serves as a model for better building design and construction, rehabilitation and maintenance, and product development around the world.

GOALS TO ACHIEVE THE VISION

Our Objective: To Achieve Major Human Health Gains Over the Next 50 Years By Upgrading Indoor Environments.

GOAL 1:

Achieve Major Health Gains and Improve Professional Education

- Known risks from indoor environments are effectively addressed, leading to significant health gains in many areas, including:
  - Avoidance of excess lung cancer deaths caused by exposure to radon, ETS, and asbestos.
  - Avoidance of excess cancer deaths caused by indoor exposure to volatile organic compounds and other chemicals.
  - Avoidance of delays in physical and mental development in children, lowered IQ levels, shortened attention spans, and behavioral problems associated with elevated blood lead levels.
  - Avoidance of excess deaths, illness, and lost school/work time from asthma and other respiratory diseases; improved comfort for the estimated 17 million Americans who have asthma, and many of the more than 50 million who suffer from allergies.
  - Significant reductions in the spread of infectious diseases, such as tuberculosis, Legionnaires' disease, and influenza
  - Significant reductions in other health effects, including eye, nose and throat irritation, headaches, fatigue, loss of coordination, nausea, developmental and reproductive damage, and damage to the liver, kidneys, and central nervous system.
  - Major productivity gains from improvements in worker and student performance.
- The most important risks posed by indoor environments are identified and quantified, and communicated in an appropriate manner for the general public; risks from interactions of toxins and cumulative low-level exposures are clarified.

GOAL 1:
Achieve Major Health Gains and Improve Professional Education

GOAL 2:
Foster a Revolution in the Design of New and Renovated Buildings

GOAL 3:
Stimulate Nationwide Action to Enhance Health in Existing Structures

GOAL 4:
Create and Use Innovative Products, Materials, and Technologies

GOAL 5:
Promote Health-Conscious Individual Behavior and Consumer Awareness

- Information about indoor health risks and healthy indoor environments is fully integrated into professional curricula and the training and practice of health, science, manufacturing, and building professionals.

GOAL 2:

Foster a Revolution in the Design of New and Renovated Buildings

- The planning and construction of nearly all new buildings is based on an integrated design[1] process that looks at whole-building systems and seeks out least-cost strategies for simultaneously achieving health-enhancing indoor environments, efficiency in the use of energy and materials, high functionality, comfort, and productivity.
- New buildings with health-enhancing indoor environments and high energy efficiency cost no more to own and operate than conventional buildings, and often cost less.
- Cost-effective building rehabilitation, renovation, and remodeling strategies have been developed that address all major identified risks in the indoor environment.
- Residential and non-residential rehabilitation, renovation, and remodeling projects are undertaken using integrated design processes and achieve large improvements in indoor environmental quality, energy efficiency, and productivity.
- The great majority of new and renovated buildings are designed for easy maintenance with low-impact products and procedures.

GOAL 3:

Stimulate Nationwide Action to Enhance Health in Existing Structures

- Guidelines for healthy building operation, maintenance, renovation, and remodeling are developed and routinely followed in commercial and office buildings.
- Standards of care and livability are developed and routinely followed in residential rental buildings.
- Building managers and engineers, maintenance and custodial workers, trash handlers, recyclers, renovators, and others who contribute directly to maintaining clean, healthy indoor environments have proper training and the capability to carry out their work.

GOAL 4:

Create and Use Innovative Products, Materials, and Technologies

- Building materials and consumer products that pose potential health and environmental risks are subject to standardized, life-simulation tests
- An easily understood "green labeling" system has been developed that allows consumers to assess health risks and make informed choices among building materials and consumer products used indoors.
- Low-toxicity, resource-efficient products are widely available in all indoor product areas and usually cost no more than conventional products.
- Low-cost testing kits and sensors for detecting exposure to a wide range of indoor pollutants and assessing personal risks are available to all.

GOAL 5:

Promote Health-Conscious Individual Behavior and Consumer Awareness

- Everyone is aware of the importance of indoor environments for maintaining and enhancing health. Indoor environmental quality is seen as just as important to health as environmental quality outdoors.
- Nearly everyone is familiar with how to access information about indoor environments, including information on health effects, environmental impacts, pollution prevention strategies, integrated building design, and indoor enviro-friendly products.
- Information is easily available and useful to the general public (including access in multiple languages) and to all relevant constituencies, including building professionals, product manufacturers, and health professionals.
- Nearly all involuntary exposure to ETS has ceased, so that it is no longer a significant health threat.
- The great majority of homeowners, tenants, and landlords significantly reduce exposure to indoor pollutants and irritants by practicing good building maintenance, e.g., controlling moisture problems, exterminating cockroaches and other vermin, changing air filters, and testing for radon, lead, asbestos, carbon monoxide, and other toxics.
- More informed consumer product purchasing and use has led to a substantial reduction in health risks associated with cleaning, painting, lawn and garden care, and other aspects of personal behavior in home environments.

PRINCIPLES FOR THE FUTURE

Whole-Systems Thinking

From the smallest home to the largest office building, we will improve indoor environments through an integrated design process that looks at the building life cycle, whole-building systems, and the leverage points where individual expenditures can generate multiple benefits. A whole-systems perspective that evaluates the building, including its site, heating, ventilating, and air-conditioning (HVAC) systems, materials, finishes, carpets, paints, appliances, and equipment, is the key to implementing the least-cost design and remodeling.

Protecting Our Children

We will create indoor environments that are healthier for everyone by making indoor environments safer for the most vulnerable among us, especially children. Improving indoor environments is critical to children's health and lays the groundwork for healthier generations to come.

Environmental Justice

Some population groups—usually low-income people, and often minorities—are exposed to a disproportional amount of environmental hazards both indoors and outdoors, at home and at work. Economically disadvantaged people often have fewer chances to improve their housing or workplace conditions. Environmental justice requires that we make extra efforts to ensure that these groups are equally protected.

Right-To-Know

Citizens have a right to know what is in their environment and how it affects them. This allows them to make informed choices to protect themselves from environmental health threats. This principle applies to indoor environments just as much as to outdoor ones. It will become increasingly relevant as low-emission products and low-cost indoor testing and sensor technologies are marketed.

ENHANCING HEALTH

Indoor environmental conditions can lead people to under-perform and to feel less than their best without producing overt symptoms of illness. Efforts to improve indoor environments should not only prevent illness, but also aim to enhance health, vitality, and productive activity.

GOOD SCIENCE

Strategies for improving indoor environments need to be based on scientific facts. We need a better understanding of indoor sources, people's exposure in various indoor environments, how those exposures affect health and productivity, and how they can be minimized by prevention-oriented, least-cost strategies. We also need to know more about the relationships between pollutant levels and building characteristics, operation and maintenance, and furnishings.

POLLUTION PREVENTION/HEALTH PROMOTION

Preventing indoor environmental problems from occurring in the first place is far more cost-effective than remediating problems and treating illnesses after they occur. The key to stopping the escalation of health care costs is to "design out" the conditions which cause illness, including unhealthy environments in homes, schools, and workplaces.

IMPORTANCE OF WIDESPREAD UNDERSTANDING

Indoor environmental quality is the sum total of decisions made by an enormous variety of individuals and institutions, including architects and builders; bankers and real estate agents; academic scientists and medical professionals; national, state, and local governments; building owners and managers; product manufacturers and retailers; janitors and sanitation workers; employers and unions; parents; consumers; and others. We can improve indoor environments faster if all these parties become more knowledgeable, so that the impetus for change emerges from all sectors.

Creative Partnerships

Improving human health indoors requires a new level of systematic cooperation among disciplines, as well as the many public, private, and voluntary organizations whose activity affects indoor environments. Partnerships among disciplines are critical for whole-systems thinking and integrated design. Partnerships between government and business can bolster research efforts and speed the emergence of profitable solutions. Partnerships among federal, state, and local governments can accomplish far more than federal action alone. Such creative linkages are key to improving indoor environmental quality and lowering health risks.

Federal Integration and Leadership

Improving indoor environments requires better coordination within and among federal government agencies to align efforts and set clear roles for each organization. New efforts will be required within EPA to span internal boundaries and collaborate more effectively with other agencies.

The federal government can lead by example and through implementing strategies that empower others. Key areas for federal action include constructing facilities that are models of IEQ and developing criteria and management systems for IEQ that provide examples for other agencies. (EPA specifically considered IEQ as it designed and constructed its new research facility in Research Triangle Park, NC and continues to consider the impacts on IEQ as it makes decisions about products and materials to be used or installed in the building.) The federal government can also enable other stakeholders, using strategies such as supporting research and development, providing information, stimulating the marketplace through purchasing, and setting standards or encouraging efforts to develop standards within the private sector.

Resource and Economic Efficiency

Strategies for improving indoor environments can be designed to promote economic efficiency, spur technological innovation, and benefit business, while promoting public health at the same time. Choosing designs, materials, and products wisely will create healthy indoor environments while simultaneously improving efficient use of energy and materials. Energy use can be reduced through more efficient building envelopes, glazings, and lighting systems. The need for materials can be reduced by minimum-materials design, minimum-toxicity components, improved durability, more flexible building design (so buildings do not have to be replaced when their use changes), and more extensive recycling and reuse of building materials. Technologies to use water more efficiently can also play an important role in areas where water supplies are limited. Money can be saved by downsizing HVAC equipment, reducing material costs, and cutting operating expenses for heating, cooling, and lighting. Small businesses may require special strategies to enable them to stay competitive and improve indoor environments.

Interrelationship of Indoor and Outdoor Environments

We need to take into account the relationship between indoor and outdoor environments. Some health endpoints, like asthma, are impacted by both, and the contribution of each is not separable. Designs for healthy buildings should include energy-saving landscaping and should be tailored to deal with the special problems posed by regional climates and local conditions. Healthy indoor environments are easier to achieve when outdoor environmental quality is high, because what comes in from the outside affects the indoors.

Goals and Measurements

Clear goals and measurements of movement toward them must be set. Appropriate measurements of success are essential for tracking and demonstrating progress, evaluating programs, and directing strategies.

END NOTE

[1]Integrated design looks at all the parameters of a building, including its site, over its lifetime, and finds the maximum balance between good IEQ, energy efficiency, comfort, and materials use.

CHAPTER 3

POTENTIAL ACTIONS

OVERVIEW

The following is an outline of potential actions for the five goals identified in Chapter 2.

GOAL 1:

Achieve Major Health Gains and Improve Professional Education

A. Develop a risk assessment methodology, perform research, and conduct assessments.

B. Along with other public health agencies, develop a public health metric (or series of metrics) as a baseline against which to demonstrate health gains.

C. Demonstrate specific health gains from good IEQ practices and marshal evidence to indicate that the gains are due to actions taken.

D. Provide information/education to foster understanding and action.

GOAL 2:

Foster a Revolution in the Design of New and Renovated Buildings

A. Quantify the benefits and costs of integrated design and use this information to provide incentives to build/renovate buildings with integrated building designs.

B. Facilitate competitions or industry consortia to develop integrated building designs.

C. Develop and promote building system performance targets.

D. Develop university and continuing education curricula.

Goal 3:

Stimulate Nationwide Action to Enhance Health in Existing Structures

A. Identify and fill knowledge gaps for the full range of existing buildings.

B. Develop and promote excellent IEQ standards of care.

C. Develop specific guidance documents for critical junctures in the life cycle of existing buildings.

D. Develop metrics for a performance-based building rating/ certification program.

E. Provide information targeted to do-it-yourselfers.

F. Develop homeowner/tenant checklists.

Goal 4:

Create and Use Innovative Products, Materials, and Technologies

A. Further develop tools to prioritize our efforts to reduce risks from sources and pollutants indoors.

B. Document and evaluate state-of-the-art sensors, test kits, and indoor-related prevention and control technologies.

C. Perform comparative exposure and risk assessments on products and materials.

D. Develop product testing protocols.

E. Work with stakeholders and outside standard-setting organizations to develop voluntary, consensus-based standards and guidelines.

F. Provide market incentives to drive manufacturers to develop both new products and new technologies.

G. Work with interested stakeholders to develop and disseminate product labels, instructional materials, enhanced material safety data sheets, and product specifications.

GOAL 5:

Promote Health-Conscious Individual Behavior and Consumer Awareness

A. Initiate a campaign to educate society's leaders on IEQ.

B. Create a healthy children program.

C. Ensure consumers are well-informed.

D. Provide for healthy home care.

POTENTIAL ACTIONS FOR GOAL 1

> **GOAL 1**
> Achieve Major Health Gains and Improve Professional Education

A fundamental requirement for improving human health indoors is a better understanding of the health risks posed by different types of indoor environments. A comprehensive assessment of health risks across the wide variety of indoor environments, and their relationship to ambient pollutants, will require extensive research efforts. Issues needing further research include test methods, basic toxicology for agents and mixtures, and the development of biomarkers and appropriate environmental measurements. A sustained, long-term effort is needed to identify and quantify the most important indoor health risks.

To demonstrate how healthier buildings lead to healthier people, research is also needed to establish public health baselines against which health gains can be measured. Critical to achieving this goal is the quick communication of research findings about indoor health risks, and how they can be avoided, to building and public health professionals, product manufacturers, and the public. Metrics are needed to measure the status and trends of a number of health effects caused by poor indoor environmental quality. This effort will require coordination with other public health agencies interested in indoor environmental issues. Once metrics are established, they can be used to demonstrate health gains from appropriate risk management options.

We can improve the indoor environment most rapidly if all parties involved become more knowledgeable, so that the impetus for change comes from all directions.

A. Develop a risk assessment methodology, perform research, and conduct assessments.

   These assessments will determine how potential risks posed by indoor exposures can be predicted accurately, quickly, and cost-effectively

1. Address multiple pathways, multiple agents, and non-traditional stressors (e.g., thermal, light, sound).
2. Develop a peer-reviewed, high-level cross-Agency research strategy, with buy-in from other agencies as well as non-federal stakeholders, designed to improve public health. This strategy may be developed at the level of the White House Committee on Environment and Natural Resources and should address:
   - Appropriate test methods to assess the often symptom/complaint-related issues associated with indoor environments.
   - Toxicological testing for agents and mixtures, especially for agents affecting immunotoxicity, neurotoxicity, and human performance.
   - Vulnerable populations, particularly children.
   - Development of biomarkers and appropriate environmental measurements.
   - A testing strategy to help address associated health risks.
   - Measurements and models to determine exposures in various indoor microenvironments.
   - Methods and models to quantify emissions from indoor sources and determine penetration of ambient pollutants indoors.
   - Innovative risk management options to reduce exposures.
3. Manage a coordinated effort (government and non-government) to perform the necessary exposure, effects assessment, and risk management research.
4. Complete the EPA portion of the inter-agency research effort.
5. Establish an indoor environmental risk assessment methodology and databases for ready access.

B. Along with other public health agencies, develop a public health metric (or series of metrics) as a baseline against which to demonstrate health gains.

Metrics are needed to measure status and trends for asthma and allergens, productivity/human performance, irritancy, neurotoxicity, reproductive toxicity, infectious disease, cancer, and other health impacts.

1. Identify health conditions that should be included in the public health baseline.

2. Ensure:

   - Collection of the necessary public health data to assess the public health baseline.
   - Development and acceptance of public health indicators and metrics.
   - A commitment to using the indicators and metrics on a national scale.

C. Demonstrate specific health gains from good IEQ practices and marshal evidence to indicate that the gains are due to actions taken

1. Identify specific actions, and ensure that the actions are implemented, documented, and tracked.

2. Push aggressively to implement those actions likely to produce the largest reduction for each known risk.

3. Monitor national status and trends of public health by working with other public health agencies.

4. Demonstrate the link between improved public health and actions taken by assessing changes to the public health baseline.

D. Provide information/education to foster understanding and action.

1. Integrate information about indoor health risks and healthy indoor environments into professional curricula and health professional training, as well as training of building professionals.

   a. Include case studies in the educational curricula of medical professionals, architects, and engineers.

   b. Educate insurance and real estate agents, building sanitation engineers, code enforcement and code writing bodies, mortgage lenders, etc.

2. Develop health issue papers for the public on such known risks as:

   - Radon
   - Environmental tobacco smoke
   - Lead poisoning
   - Asthma/allergies
   - Infectious diseases
   - Reduced productivity from symptom-based conditions

3. Develop additional health issue papers as research identifies further hazards.

POTENTIAL ACTIONS FOR GOAL 2

GOAL 2

Foster a Revolution in the Design of New and Renovated Buildings

Dramatic improvements in the indoor environments of the next century will be achieved with integrated design and good indoor environmental quality planning and construction. The design and construction of new residential and commercial buildings accounts for some $381 billion per year in the U.S. economy, with new homes alone accounting for $182 billion (U.S. DOC 1996, 1997). Extensive building renovations offer similar opportunities for improving indoor environments. Once the poor stepsister of the building industry, non-residential rehabilitation is now a major market. Most of this work requires total building overhaul or major renovations, not just remodeling or repair.

A number of buildings have been constructed around the world over the past few years using an integrated design process. They demonstrate that improvements in energy efficiency can complement indoor environmental upgrades. Often, integrated design actually saves money by downsizing HVAC equipment, reducing material costs, and cutting operating expenses for heating, cooling, and lighting. New and renovated buildings can also be designed for easy maintenance with low-impact, high-efficiency products and procedures.

Several kinds of initiatives are needed to help integrated design move from its present status of "innovative best practice" to standard practice. Research is needed to establish the economic costs and benefits of integrated design and good IEQ construction, as well as the costs of health care, productivity loss, and poor building performance related to inferior ventilation, IEQ design, and construction. Reliable information of this kind will eventually influence costs for insurance, mortgages, and health care coverage, creating strong economic incentives for integrated design.

Needed as well are new tools to provide industry and consumers with the information they require to make sound building and renovation decisions. Professionals in the design and building industries need to agree on the elements of good IEQ design and on appropriate ways to measure and compare the features offered by given designs. Collaboration with professionals and organizations in the design, engineering, construction, building products, real estate, government, and public health communities is essential to speed change in professional practice, professional curricula, and standards and code setting.

A. Quantify the benefits and costs of integrated design and use this information to provide incentives to build/renovate buildings with integrated building designs.

Integrated design simultaneously achieves good indoor environmental quality, energy efficiency, high functionality, comfort, and productivity. Building design, construction, and procurement professionals need sound financial arguments to make healthy indoor environment features a priority in new buildings.

1. Convene a stakeholder process to define good/superior IEQ for various building types.

2. Collect existing information and perform needed research to quantify initial building and lifetime costs of superior IEQ and the savings from improved health, productivity, and building systems performance. Focus on energy, productivity, absenteeism, cost of law suits and worker's compensation, tenant turnover/retention, sale of homes, rental rates, costs of implementing guidance, and assessment of the market value for a "healthy building."

3. Use existing data and research results to develop building design simulation packages that demonstrate the consequences of building design and product choice on the health, economics, and productivity of the occupants. Address air quality, air flow, energy consumption, life cycle effects, moisture intrusion, humidity, and health/productivity impacts.

4. Promote integrated design cost/benefit information for good decision making by:

   a. Widely disseminating cost/benefit information to builders, product manufacturers, commercial realtors, insurance and mortgage companies, public health professionals, and consumers to improve understanding and encourage integrated designs.

b. Championing insurance industry rate incentives for superior IEQ buildings using cost-benefit arguments. Work with consumer advocacy organizations and insurance companies or their professional organizations to pioneer reduced premium costs for holders of health, home, or commercial property insurance policies who have created high quality indoor environments.

c. Creating primary and secondary mortgage banking instruments that result in savings for residential remodeling and new construction projects that use integrated design. Use cost-benefit arguments that demonstrate savings from improved systems performance, lowered taxes, and improved insurance rates, and work with consumer groups and mortgage bankers to craft lower debt-to-equity rates for residential lending. This program may be modeled on or integrated with the Energy Efficient Mortgage program.

d. Working with school districts to revise how schools allocate resources, taking into account the cost of new construction, maintenance, and future costs, and revising federal and state formulas to reflect these factors.

e. Educating consumers on the benefits of an integrated design approach.

B. Facilitate competitions or industry consortia to develop integrated building designs.

Options to be considered:

1. Establish a consortia of designers, manufacturers, and other stakeholders to develop the designs and the building materials for high-performance buildings.

2. Promote juried design competitions, undertaken with other stakeholders, that focus the creativity of architects and designers on improved indoor environments.

3. Provide grants to show that integrated designs are feasible.

C. Develop and promote building system performance targets.

1. Through a stakeholder process, develop IEQ performance targets for new or renovated buildings. Work with established voluntary standards-setting organizations to create a unified set of voluntary standards, incorporating key IEQ-related variables (maintainability, air quality, energy efficiency, air flow, ventilation, materials selection, limiting moisture intrusion, controlling humidity, and feedback loops for measurement and evaluation). Develop a voluntary ratings system that predicts performance.

2. Procure a Presidential Executive Order requiring new or renovated federal buildings to comply with the voluntary standards.

3. Establish a green codes program where localities lower permitting fees, cut taxes, or simplify procedures for buildings that adhere to a voluntary IEQ buildings rating system. Work with international code officials and local government organizations responsible for building codes to develop model programs. Where possible, integrate green code efforts with American Institute of Architects (AIA) and other existing green buildings efforts (U.S. Green Buildings Council, Energy Star®, the Office of Policy, Economics, and Innovation (OPEI) Smart Growth Network, and the OPEI/Office of Solid Waste (OSW)/National Association of Home Builders (NAHB) Research Center program) to develop a local green builder program model.

4. Create a recognition program for integrated-design buildings. Highlight buildings built with whole-systems design, including

schools, office buildings, and low- and moderate-income housing. Assist in developing an industry organization or an independent authority to establish and oversee a recognition program that directs potential consumers to the benefits offered by the building's good IEQ. Integrate these efforts with existing green buildings efforts.

D. Develop university and continuing education curricula.

Develop curricula and work with state licensing agencies to incorporate integrated design standards and continuing education requirements for designers, architects, engineers, and health professionals. Begin a certification process for companies and individuals for added marketability and develop a mechanism for recognition and price differentiation in the marketplace. Develop integrated design components for existing professional training programs. Partner with state contractor licensing organizations, builders, remodelers, and other industry groups to promote integrated design standards.

POTENTIAL ACTIONS FOR GOAL 3

GOAL 3

Stimulate Nationwide Action to Enhance Health in Existing Structures

The indoor environments of existing structures must also be considered. Each year, the inventory of existing buildings grows both older and larger. From the point of view of human health, therefore, it is important to improve the indoor environmental quality of existing buildings so that virtually everyone lives and works in healthy surroundings.

Because industrial environments are unique and, for the most part, well-regulated, our efforts focus on non-industrial buildings of all types. These non-industrial buildings range from single-family, owner-occupied structures to large multi-tenanted residential buildings; from small retail establishments to large office buildings; from hospitals to prisons to schools.

Several types of initiatives can combine to improve IEQ in existing buildings. Guidelines can be developed and promoted for improving IEQ in routine remodeling and repairs. Standards of care and livability for healthy building operation and maintenance can be institutionalized. Research can support the development of guidance and make outreach programs more effective. Education and training programs can ensure that those responsible for managing and maintaining buildings have the ability to perform their work. Better measures of building performance and recognition programs can heighten awareness of the issue in general and the status of particular buildings. Enhancements to energy efficiency can be made in tandem and can often pay the bill for IEQ improvements.

A. Identify and fill knowledge gaps for the full range of existing buildings.

   Buildings of interest cover a wide spectrum and can include residences, hospitals, and hotels.

   1. Develop and carry out a data inventory analysis and a research agenda in the following areas:

      - Current IEQ in non-office buildings. (EPA has recently completed the data collection phase of a baseline study of office buildings.)
      - Short–and long–term costs and benefits of good IEQ, including such factors as improved health (and health costs), energy, productivity, absenteeism, cost of law suits and worker's compensation, tenant

turnover/retention, fire susceptibility, equipment life expectancy, sale of homes, rental rates, costs of implementing guidance, and assessment of the market value for a "healthy building."

- IEQ diagnostic protocols and detection technologies.
- Building ventilation control technologies that are most effective from IEQ and energy standpoints.
- Building maintenance protocols and their impact on IEQ, including cleaning and maintenance products.
- Building IEQ remediation protocols.

2. Target a building with stable historical data and identify stakeholders who can study the effectiveness of EPA's guidance in improving IEQ and its effect on health, including quantifying effects through health insurance claims, sick leave, and productivity gains and losses.

B. Develop and promote excellent IEQ standards of care.

1. Work with stakeholders to facilitate the creation of integrated IEQ standards of care for different building types, taking into account the interrelated roles and responsibilities of building owners, managers, occupants, and tenants.
2. Procure a Presidential Executive Order requiring existing federal buildings (both owned and leased) to comply with integrated IEQ standards of care.
3. Encourage the adoption of IEQ standards of care in mortgage and insurance policies and rates, hospital certification, voluntary practice guidelines, and building codes.
4. Develop a voluntary Building Coalition dedicated to promoting the adoption of IEQ standards of care. The Coalition could: develop an outreach mechanism/tool to encourage adoption, including the development of training outlined below; create a building recognition

program; manage the development of the IEQ performance index; and serve as the focal point for future progress on IEQ in existing buildings.

5. With stakeholders, develop training and other tools to educate various audiences on IEQ standards of care. Promote adoption by institutions that educate architects, engineers, home inspectors, and other building professionals, and as continuing education or a prerequisite for certification by professional organizations. Promote education and training programs to ensure that building managers and engineers, maintenance and custodial workers, trash handlers, pest management contractors, recyclers, and others who contribute directly to maintaining indoor environments have the information and capabilities they need for carrying out their work. Ensure that training and other tools reach other IEQ service providers, local health officials, and residential audiences.

C. Develop specific guidance documents for critical junctures in the life cycle of existing buildings.

1. Improve the indoor environment and educate people about the IEQ effects of decisions and activities that may result in increased hazards, including disturbance of asbestos and lead paint. Key events may occur:

- During remodeling
- At sale of building/home
- At building commissioning and decommissioning
- During annual safety inspection
- During tenant improvement projects
- During building recertification
- After flooding/fire/storms

2. Develop outreach programs to encourage people to take action.

D. Develop metrics for a performance-based building rating/certification program.

1. Facilitate stakeholder development of an IEQ performance metric for different building types that utilizes research done under other action items elsewhere in this plan (e.g., baseline data on IEQ, health effects data, cost/benefit information).

2. Facilitate the establishment of a performance-based rating/certification program that utilizes an IEQ performance metric and baseline data to develop a voluntary performance standard or threshold and a verification protocol. Promote this program to building owners, insurers, occupants, government officials, and consumers using a variety of success stories.

E. Provide information targeted to do-it-yourselfers.

Work in partnership with major hardware retailers, and other organizations that target do-it-yourselfers, to include point-of-purchase displays, print advertising, promotion of products meeting good IEQ standards, and homeowner on-line workshops with IEQ experts. Focus materials and activities on raising homeowner awareness of good renovation design for IEQ.

F. Develop homeowner/tenant checklists.

Develop instruments that allow homeowners/tenants to do their own IEQ self audits which identify issues in the home and stress the importance of proper cleaning and maintenance of home appliances. Develop and implement a strategy to disseminate these checklists widely.

POTENTIAL ACTIONS FOR GOAL 4

GOAL 4

Create and Use Innovative Products, Materials, and Technologies

The products, materials, and technologies that we use inside our buildings are another potential source of indoor environmental problems. A key component for achieving building improvements is the use of building materials, during construction and renovation, which produce low levels of any potentially harmful emissions.

Many strategies are available to accelerate the innovation of products, materials, and technologies. The most fundamental approach is to develop a reliable emissions testing system, to perform comparative risk assessments, and to develop voluntary, consensus-based guidelines and standards to assist in the evaluation of products, materials, and technologies. The results of standardized testing can be used to develop low-toxicity products that are competitively priced with conventional products and can serve as a basis for developing information to assist consumers in making informed choices among products, materials, and technologies used indoors.

Voluntary guidelines and standards for products, materials, and technologies can take many forms. For example, guidelines or standards might ask a manufacturer to provide emission levels from a product for comparative purposes with other similar products, or they may set a level above which a product is regarded as "unsafe." They may also ensure the appropriate use of products (e.g., labeling on the use of adequate ventilation or limiting the use to certified applicators). In addition, certain industry members who perform well in reducing emission levels may be recognized under a program similar to the EPA Green Lights® or Energy Star® programs.

While the first line of defense is to prevent pollution by controlling sources of indoor pollutants, rapid progress is also needed in monitoring and control technologies. Low-cost sensors and test kits, for example, will eventually make it possible for nearly everyone to assess their risks indoors. Improvements are needed in technologies to "clean" air and to increase ventilation efficiency in buildings.

A. Further develop tools to prioritize our efforts to reduce risks from sources and pollutants indoors.

In consultation with all interested stakeholders, continue developing and using tools to prioritize those products and materials that may present the greatest exposures and risks to human health indoors. This prioritization should address both the health risks and potential benefits of the products and materials. Continually review and update the tools as new information becomes available on product formulations and emissions, exposure data, and toxicity information.

1. Work with all interested stakeholders to collect and compile additional existing data to assist in this prioritization.

2. Seek input, through stakeholder workshops, on those priority consumer products and building materials that, based upon the best available data, have the largest relative impact on health in indoor environments based on both chemical and biological contaminant emissions.

3. In the interest of the public's right to know, make summaries of the publicly available chemical formulations in product categories available through a web page and prepare chemical exposure and toxicity fact sheets for the chemicals and product categories that are accessible through this web page.

B. Document and evaluate state-of-the-art sensors, test kits, and indoor-related prevention and control technologies.

1. Survey, monitor, document, and assess the status and progress of technologies based on efficacy, health impacts (e.g., enhanced growth of microorganisms, chemical emissions), and cost; identify the trends, technical issues, and needs of future development.

2. Publish and periodically update this analysis in a database, by technology type; use the database as a source of information on current technologies and as a measurement tool to assess progress in stimulating research and development to improve these devices.

C. Perform comparative exposure and risk assessments on products and materials.

1. Provide leadership in working with outside stakeholders to establish an exposure and health risk assessment methodology for consumer products and building materials used indoors that would address the total health impacts of products, including beneficial impacts (e.g., disinfection).

2. Develop consensus on the general methods to be used to consider relevant information, including:

   - Data on all routes of exposure (nasal, inhalation, dermal, and oral) and their comparative importance.
   - The effects of indoor sinks and interactions of multiple pollutants from multiple sources on indoor exposure levels.
   - Both toxicological and sensory health impacts.
   - Evaluation of dose-response relationships.

D. Develop product testing protocols.

1. Establish a standardized, consensus-based generalized emissions testing system with stakeholders, so that the potential exposure and health risk of most consumer products and building materials can be assessed. Develop and validate low-toxicity products using the testing and assessment system.

2. Assist stakeholders in developing standardized, consensus-based emissions testing and risk assessment systems specific to their products or materials and in promoting the concept of low-toxicity products and materials.

E. Work with stakeholders and outside standard-setting organizations to develop voluntary, consensus-based standards and guidelines.

1. Develop standards or guidelines for emissions levels of chemicals from products and materials used indoors by convening a dialogue to set consensus-based ground rules that can be used by organizations outside the federal government to develop standards and guidelines.

2. Develop guidance on safe levels of pollutants in indoor environments to assist in the development of sensor and control technologies.

3. Develop standards or guidelines to evaluate the efficacy, reliability, and cost-effectiveness of new technologies used to monitor or control pollutants (e.g., sensors, air cleaners).

F. Provide market incentives to drive manufacturers to develop both new products and new technologies.

The incentives will provide for healthier indoor environments and will not compromise other aspects of environmental performance. EPA will lead other stakeholders in working with a wide range of consumers to direct demand toward healthier indoor products and technologies.

1. Focus on creative market incentives such as those derived from financing and insurance mechanisms (e.g., discounts in health insurance rates for people who live in homes with healthier indoor environments).

2. Work with institutional buyers within the federal government and elsewhere (e.g., hospitals, schools and universities, retail sector) to increase demand for cleaner indoor products through individual pilot projects focusing on specific products or materials. Establish bidding procedures for manufacturers to compete on the basis of both price *and* emissions to ensure lower emissions at reasonable prices. Develop a database and communications program to collect

the experience and bid results from participants and to communicate information on technical feasibility and cost to spur new buyer membership and competition by manufacturers.

3. Periodically survey the market to gauge the extent to which demand rises for cleaner products and the extent to which that demand is leading towards improvements in products and technologies.

4. Develop and implement a healthy products award program to recognize companies that develop, market, or purchase cleaner indoor products.

5. Provide programmatic grants to product manufacturers and other parties to develop low-emitting or low-toxicity products that are less problematic from a public health perspective.

6. Promote IEQ-friendly products through the development of planning and sales software for building contractors. Integrate this effort with the Office of Prevention, Pesticides, and Toxic Substances' Environmentally Preferable Products program.

7. Make cost-effective IEQ monitors and control technologies a standard feature.

    a. Identify stakeholders to popularize the standard use of basic detection systems for home and work and help develop a clearinghouse for appropriate sensor and mitigation technologies.

    b. Fund an effort, possibly through programmatic demonstration grants, to integrate reliable indoor sensor technologies with environmental controls (i.e.,"smart" building systems) in institutional settings, such as offices, hospitals, schools, and prisons, as a means to create an awareness of indoor pollutants and demand for a healthier indoor environment.

c. Integrate available and reliable indoor sensor technologies with environmental control systems in residential settings to create consumer demand for healthier indoor environments.

d. Initiate a field study to evaluate commercially available indoor pollutant monitors and control devices for both performance and practicality.

e. Assure federal adoption of new systems.

C. Work with interested stakeholders to develop and disseminate product labels, instructional materials, enhanced material safety data sheets, and product specifications that will allow for the incorporation of a broad spectrum of environmental and performance information. These materials can be used by consumers to select the best products, materials, and new technologies for use indoors.

1. Encourage consumers to make informed choices when deciding what products, materials, and technologies to purchase for use indoors, as well as how they should be used.

2. Collect background research and conduct individual interviews and focus group discussions to develop specific recommendations for the type and design of user information to enable consumers to weigh environmental impacts indoors, as well as product performance and beneficial aspects, in purchasing decisions for both products and new technologies.

3. Develop appropriate user information, which focuses on reducing human health risks in the indoor environment and includes information on the beneficial aspects and performance characteristics of the products and materials. Convene all interested stakeholder groups, including industry, institutional purchasers, and organizations experienced in providing user information for priority indoor products, technologies, and services.

POTENTIAL ACTIONS FOR GOAL 5

GOAL 5

Promote Health-Conscious Individual Behavior and Consumer Awareness

More health-conscious individual behavior can create healthier indoor environments. In an ideal situation, nearly everyone sees indoor environmental quality as important for health and most people know how to get information they need. For individuals to engage in health-conscious behavior regarding their indoor environment, they must be informed, have the tools necessary to act, and believe their actions will result in a benefit to their health, lifestyle, or productivity.

Improving indoor environmental quality and reducing the health risks of serious indoor environmental problems will require millions of self-initiated actions by individual home dwellers, building owners and managers, parents, school officials, real estate professionals, and other key target audiences. Effective programs to achieve this mission must emphasize communication and outreach to catalyze and influence actions by the millions of individuals who make decisions affecting indoor environments.

The following list of specific recommended initiatives uses a variety of targeted approaches for encouraging health-conscious individual behaviors to improve the indoor environment. As further research into indoor environmental health risks and mitigation strategies is conducted, new initiatives to encourage health-conscious individual behaviors will be developed.

A. Initiate a campaign to educate society's leaders on IEQ.

   1. Work with private sector leaders and public policy makers at the federal, state, and local levels to demonstrate the significance of the indoor environment and the cost-effective benefits of improved conditions in homes, schools, workplaces, and public buildings.

   2. Develop a highly-targeted campaign aimed at encouraging society's leaders to understand the following key facts about indoor environmental quality:

      - People spend 90 percent of their time indoors.
      - Indoor environmental problems are high risk.
      - There are cost-effective solutions to many IEQ problems.

- Research is needed to improve our understanding of how to prevent IEQ problems.

3. Reach out to scientists, influential medical centers, high-level health officials, state legislators, tribal leaders, private sector executives, influential state and local officials, and other key opinion leaders. Use a variety of targeted channels ranging from scientific journals to the mass media, including articles in popular publications and airline flight magazines, speakers at key conventions, and feature segments in TV programs and Sunday morning talk shows. Conduct these activities in partnership with key stakeholders.

B. Create a healthy children program.

1. Protect children from asthma by reducing the degree to which indoor environmental conditions contribute to the rate and severity of asthma in children. Work in close partnership with other federal agencies to: integrate prevention messages into existing treatment messages; emphasize innovative outreach in homes; use schools to deliver proven asthma prevention and management messages to children of preschool and primary school age; track the effectiveness of these school interventions; leverage the existing health care system to reduce costs by promoting asthma prevention and management education; and employ cutting-edge mass media approaches to raise parent and child awareness and induce health-promoting behavioral changes.

2. Develop an action campaign to improve the indoor environments of children. Form a cross-government team, including EPA representatives from OPPTS, OAR, and the Office of Children's Health Protection (OCHP), to improve the indoor environments of children in homes, day care facilities, and schools. Work with stakeholders to educate parents, day care providers, child health care providers, and school officials on the benefits of reducing children's exposure to lead, secondhand smoke, radon, allergens, pesticides, and other harmful indoor pollutants. Explore partnerships with health maintenance

organizations (HMOs) to encourage participating physicians to include environmental factors in checkups. Explore mechanisms for incorporating environmental factor training into medical school programs for patient background, screening, and diagnosis.

3. Initiate a three- to five-year campaign to reduce minority children's exposure to indoor environmental tobacco smoke using transit and other media appropriate to minority audiences. Expand existing media campaigns to include TV, radio, print, transit, billboard, and other materials targeted specifically to minority populations.

4. Educate children on indoor environmental risks by teaming with stakeholders to develop curricula, science lessons, teaching modules, and other mechanisms for mainstreaming indoor environmental subject matter into the Nation's formal education system. Teaching children about the importance of the indoor environment to human health will help to ensure health-conscious behaviors in two long-term ways: (1) by developing an awareness of how the indoor environment impacts health and productivity so that children will ultimately be better managers of their own indoor environments as adults and (2) when children adopt environmentally conscious behaviors, the adults in their lives often emulate those behaviors (e.g., recycling).

C. Ensure consumers are well-informed.

1. Take a comprehensive approach to the real estate sector, which provides a critical link to achieving measurable risk reduction on radon, carbon monoxide, lead in paint, asbestos, underground storage tanks, and drinking water. Agents, brokers, home inspectors, attorneys, mortgage bankers, and other real estate professionals are uniquely positioned to assist consumers in making informed decisions about correcting environmental problems before they purchase commercial and residential properties. Collaborate within EPA to develop and implement a cross-Agency strategy and workgroup, integrated public information materials, information clearinghouse, web site, one-stop environmental real estate hotline, and outreach partnerships with each of the major seg-

ments of the real estate professions. Engage other federal institutions (e.g., the Department of Housing and Urban Development, the Veteran's Administration (VA), Fannie Mae, Freddie Mac) to coordinate environmental requirements.

2. Publish "50 Things You Can Do to Improve Your Indoor Environment." Develop and promote clear and consistent messages on indoor environmental concerns and questions frequently asked by the public. Prepare and distribute these as concise, easy-to-use materials in multiple formats (web page, consumer advice booklet, magazine article) which clearly explain what people can do now to improve their indoor environments.

3. Encourage more informed consumer product purchasing. Engage the private sector and other concerned federal agencies in designing ways to educate consumers about how to purchase products wisely and use them with appropriate care. Consumers infrequently read product labels before using the contents and often disregard important manufacturer's instructions concerning safe use of the product. Likewise, product labels lack uniformity in the way safety and use instructions are presented. Directions such as "use with adequate ventilation" are subject to broad interpretation.

4. Initiate a consumer campaign to improve indoor workplace environments. With groups like the Occupational Safety and Health Administration (OSHA) and organized labor, develop a comprehensive information campaign to educate the public about the straightforward, cost-effective actions that can be taken to improve indoor air quality in workplaces. Adjuncts to the campaign could include a toll-free hotline number, web site, or other places where building occupants, as well as owners and operators, can receive information and resource materials.

D. Provide for healthy home care.

1. Encourage broad-based public information programs and campaigns on household cleaning and maintenance that combat indoor environmental hazards. For example, expand the Master

Home Environmentalist (MHE) Program nationwide. A small pilot program that has successfully demonstrated a change in behavior, the Master Home Environmentalist Program is a hands-on, tuition-free program that teaches people about the indoor environment in return for their commitment to teach others. Topics include ways to reduce tracking soil containing lead and pesticides into the home; proper vacuuming techniques and how to evaluate the effectiveness of vacuum cleaners; safe methods to dispose of household waste; ways to identify and fix problems related to moisture indoors; and ways to reduce bioaerosols, dust mites, bacteria, and fungi indoors.

2. Make accurate information available to the public on air cleaning and filtration equipment. Working with public and private sector stakeholders, ensure that accurate information is available to the public so consumers can make wise choices when considering air cleaning and filtration equipment. Establish a system to prevent false advertising of indoor air cleaning devices, and design a means of assessing the safety and effectiveness of new devices.

3. Establish an educational mini-grant program on moisture control and the use of microbe-resistant building materials, especially for low-income populations in high-humidity regions. Coordinate with existing educational programs on moisture-related illnesses such as asthma and Legionnaires' disease.

REFERENCES

Cobb, N., and R.A. Etzel. 1991. Unintentional carbon monoxide related deaths in the United States, 1979 through 1988. Journal of the American Medical Association 266: 659-663.

Howard, G., L.E. Wagenknecht, G.L. Burke, A. Diez-Roux, G.W. Evans, P. McGovern, F. J. Nieto, and G.S. Tell. 1998. Cigarette smoking and progression of atherosclerosis. Journal of the American Medical Association 279:119-124 (abst.).

Mannino, D.M., D.M. Homa, C.A. Pertowski, A. Ashizawa, L.L. Nixon, C.A. Johnson, L.B. Ball, E. Jack, and D.S. Kang. 1998. Surveillance for asthma–United States, 1960-1995. Morbidity and Mortality Weekly Report 47(SS-1):1-28.

National Academy of Sciences (NAS). 2000. Clearing the Air: Asthma and Indoor Air Exposures. Prepared by the Committee on the Assessment of Asthma and Indoor Air Quality, Institute of Medicine. Washington, DC: National Academy Press.

National Academy of Sciences (NAS). 1998. Health Effects of Exposure to Radon (BEIR VI). Prepared by the Committee on the Biological Effects of Ionizing Radiation (BEIR) of the National Research Council. Washington, DC: National Academy Press.

National Cancer Institute (NCI). 1999. Health Effects of Exposure to Environmental Tobacco Smoke: The Report of the California Environmental Protection Agency. Smoking and Tobacco Control Monograph No. 10. Bethesda, MD: National Cancer Institute, National Institutes of Health, U.S. Department of Health and Human Services. NIH Publication No. 99-4645.

Pope, A.M., R. Patterson, and H. Burge, eds. 1993. Indoor Allergens: Assessing and Controlling Adverse Health Effects. Washington, DC: National Academy Press.

Presidential and Congressional Commission on Risk Assessment and Risk Management. 1997a. Risk Assessment and Risk Management in Regulatory Decision-Making. Final Report, Volume 2. Washington, DC. U.S. Government Printing Office. GPO 055-000-0568-1. http://www.riskworld.com/riskcommission (accessed March 2, 2001).

Presidential and Congressional Commission on Risk Assessment and Risk Management. 1997b. Presidential/Congressional Commission Issues 71 Recommendations to Improve EPA, FDA, Other Federal Agency Approaches to Environment, Public Health Threats. Press Release, March 7, 1997. http://www.riskworld.com/Nreports/1997/risk-rpt/html/pr30797.htm (accessed March 2, 2001).

Weiss, K.B., P.J. Gergen, and T.A. Hodgson. 1992. An economic evaluation of asthma in the United States. N Engl J Med 326: 862–866.

U.S. Consumer Product Safety Commission (CPSC). 1997. CPSC Urges Annual Fuel-Burning Appliance Inspection to Prevent Deaths, Fires. U.S. Consumer Product Safety Commission Press Release #97–191. Washington, D.C.

U.S. Department of Commerce (DOC). 1997. Highlights from the Expenditures for Residential Improvements and Repairs Report. Bureau of the Census, U.S. Department of Commerce Press Release, August 4.

U.S. Department of Commerce (DOC). 1996. (C Series, C-30) Current Construction Reports.

Bureau of the Census, U.S. Department of Commerce.

U.S. Department of Health and Human Services (DHHS). 2000. Blood lead levels in young children – United States and Selected States, 1996-1999. Morbidity and Mortality Weekly Report 49(50): 1133-7. http://www.cdc.gov/mmwr/preview/mmwrhtml/mm4950a3.htm (accessed May 11, 2000).

U.S. Department of Health and Human Services (DHHS). 1997a. Update: Blood lead levels - United States, 1991–1994. Morbidity and Mortality Weekly Report 46(7): 141–146. (Correction in Volume 46, No. 26).

U.S. Department of Health and Human Services (DHHS). 1997b. Legionnellosis: Legionnaires' disease and Pontiac fever. Atlanta, GA: Centers for Disease Control and Prevention. 06:50:52.

U.S. Department of Housing and Urban Development (HUD). HUD, CPSC, and DOH Promote Poison Prevention and Consumer Safety Drive in Local HUD-assisted Housing Developments, Press Release, New York, N.Y.

U.S. Environmental Protection Agency (EPA). 1999. Clear Your Home of Asthma Triggers: Your Children Will Breathe Easier. Washington, DC: U.S. Environmental Protection Agency. EPA/402-F-99–005.

U.S. Environmental Protection Agency (EPA). 1998. A Comparison of Indoor and Outdoor Concentrations of Hazardous Air Pollutants. In: Inside IAQ, Spring/Summer 1998, pp. 1–7. Prepared by Office of Research and Development, U.S. Environmental Protection Agency, Research Triangle Park, NC. EPA/600/N-98/002.

U.S. Environmental Protection Agency (EPA). 1996. Environmental Health Threats to Children. Washington, DC: U.S. Environmental Protection Agency. EPA 175-F-96-001.

U.S. Environmental Protection Agency (EPA). 1995. The Inside Story: A Guide to Indoor Air Quality. Washington, DC: U.S. Environmental Protection Agency. EPA 402-K-93-007 (GPO 055-000-00502-8).

U.S. Environmental Protection Agency (EPA). 1992. Respiratory Health Effects of Passive Smoking: Lung Cancer and Other Disorders. Washington, DC: U.S. Environmental Protection Agency. EPA/600/6-90/006F (NTIS PB 93-134419).

U.S. Environmental Protection Agency (EPA). 1991a. Guidelines for Developmental Toxicity Assessment. Washington, DC: U.S. Environmental Protection Agency.

U.S. Environmental Protection Agency (EPA). 1991b. Indoor Air Facts No. 4 (Revised): Sick Building Syndrome. Washington, DC: U.S. Environmental Protection Agency. 402-F-94-004.

U.S. Environmental Protection Agency (EPA). 1990. Reducing Risk: Setting Priorities and Strategies for Environmental Protection. Washington, DC: U.S. Environmental Protection Agency. EPA-SAB-EC-90-021.

U.S. Environmental Protection Agency (EPA). 1987. Unfinished Business: A Comparative Assessment of Environmental Problems. Washington, DC: U.S. Environmental Protection Agency.

U.S. Environmental Protection Agency (EPA), American Lung Association, U.S. Consumer Product Safety Commission, and American Medical Association. 1994. Indoor Air Pollution: An Introduction for Health Professionals. Washington, DC: U.S. Environmental Protection Agency. EPA 402-R-94-007; GPO 1994-523-217/81322.

Appendix A

INDOOR ENVIRONMENTS: CURRENT PROGRAM PRIORITIES

OVERVIEW

EPA OFFICES WORK TOGETHER

THE CARPET POLICY DIALOGUE

This voluntary dialogue was jointly led by OPPTS and OAR and included representatives of a number of other offices within the Agency, as well as all interested stakeholders (i.e., industry, unions, public interest groups, other federal agencies). The dialogue resulted in industry agreement to test new carpet floor-covering materials for total volatile organic compound (VOC) emissions and to explore ways to lower VOC emissions from carpet products. Most importantly, the industry undertook an extensive consumer education program, in cooperation with other dialogue participants, designed to provide the public with information on the role that carpet products play in indoor air quality and ways in which consumers can make informed purchase decisions.

RADON IN DRINKING WATER

OAR is collaborating with the Office of Ground Water and Drinking Water (OGWDW) to develop a unique and innovative drinking water rule for radon. The cost-effectiveness of reducing radon risk is substantially greater for indoor air (from soil gas) than from drinking water. Because of this, EPA, in proposing a maximum contaminant level (MCL) for drinking water (64 FR 59245, November 2, 1999), also made available a higher alternative maximum contaminant level (AMCL) accompanied by a multi-media mitigation program to address risks in indoor air. The proposed regulations will provide states flexibility in how to limit the public's exposure to radon.

EPA's Strategic Plan includes program priorities aimed directly at protecting human health indoors, as well as protecting it as part of broader environmental protection programs. EPA's indoor environments programs address well-known risks, such as radon, lead, asbestos, and environmental tobacco smoke. These programs also provide tools and guidance on good indoor environmental practices in residences, schools, and office buildings. Other EPA programs are broader in scope (e.g., providing safer chemicals and products, reducing exposures to hazardous waste streams, reducing risks to disadvantaged and disproportionately exposed populations), but have the protection of human health indoors as a program component. Although EPA has made significant progress in reducing risks from some well-known hazards indoors, much remains to be done.

EPA's strategic focus revolves around four main areas: science and engineering, guidance and policy development, generating public action, and measuring results.

The Agency believes that both regulatory and non-regulatory approaches have value. Regulations mandate behavioral changes by industry and others to prevent exposure to toxic substances.

Non-regulatory processes are often used to mitigate unexpected risks or to mitigate risks through voluntary actions.

Preventive approaches:

EPA uses the Federal Insecticide, Fungicide, and Rodenticide Act (FIFRA) and the Toxic Substances Control Act (TSCA) to prevent hazardous pollutants from unnecessarily entering the indoor environment. These statutes require manufacturers or users to submit information to characterize the health risks a substance might pose before it can be manufactured or distributed. EPA can then direct the manufacturer to take measures to reduce exposure to the substance, such as limiting where and how much of the substance can be used, mandating labeling and use of protective equipment to ensure proper use, and requiring training of the people who use the substance. Regulation can also further restrict or even ban a substance when there is no other way to provide adequate protection. EPA works closely with industry and other stakeholders to assist them in reducing risks to workers, communities, and the environment by developing pollution prevention and waste minimization tools. EPA observations of chemical plant incidents and subsequent investigations are being brought to the attention of industry to learn from mistakes made and to further upgrade indoor/outdoor plant safety.

STRATEGIC FOCUS

Measuring Results

- Selecting appropriate environmental indicators to measure progress
- Continuous improvement and adjustment of strategies and activities for achieving risk reduction goals

Science and Engineering

- Targeting the greatest risk first
- Identifying and filling research gaps
- Enhancing understanding of the multi-factorial nature of indoor environmental quality

Guidance and Policy Development

- Developing and refining guidance using a broad-based consensus approach
- Preventing indoor pollution through source control and building management and construction
- Using a continuum of risk management approaches to control risks (information–motivation–incentives–mandates)

Generating Public Action

- Establishing partnerships to communicate guidance and promote effective, timely action
- Forging constructive alliances to leverage resources and to ensure statutory authorities are used effectively
- Designing market-based incentives to lower source emissions and providing the information necessary to make informed decisions

Approaches for existing risks:

In some cases, products or materials in the indoor environment may present a risk to human health indoors. Besides emissions from products and materials, chemical pollutants can be introduced to the indoor environment from contaminated potable water, outdoor air, soil, and other external sources. In some cases (e.g., asbestos, lead, radon), EPA's approaches to addressing these risks are, in part, specified by statutes.

In many cases, however, EPA's approach has been to obtain non-regulatory, voluntary actions by industry to address risks. The mechanism used for eliciting this voluntary approach has often been stakeholder dialogues. These dialogues may result in the development of voluntary guidelines and standards based on levels of a pollutant, source emissions, ventilation parameters, and building or maintenance practices either in lieu of, or in addition to, regulatory action. Other non-regulatory approaches that may be taken include risk communication, training, technical assistance, cooperative partnerships, community activities, and other pollution prevention activities.

Discussions of each office's priorities and activities for protecting human health indoors are provided below.

OFFICE OF AIR AND RADIATION (OAR)

OAR's goal is to ensure that, by 2005, 16 million more Americans are living or working in homes, schools, or office buildings with healthier indoor air than in 1994. To accomplish this, several measurable milestones have been established for 2005:

- To reduce lung cancer, respiratory diseases, and other health problems, 11.5 million more Americans will benefit from healthier indoor air in their homes by the:
  - Mitigation of 700,000 homes with high radon levels and the construction of one million homes with radon-resistant construction techniques.
  - Reduction of the proportion of households in which children ages six and under are regularly exposed to smoking from 27 percent in 1994 to 15 percent.
- To reduce IAQ-related illness, five percent of office buildings will be managed with IAQ practices consistent with EPA's *Building Air Quality* guidance.
- To reduce health problems in the nearly 10 million children made ill annually from indoor air problems in schools, 15 percent of the Nation's schools will adopt good IAQ practices consistent with EPA's *IAQ Tools for Schools* guidance.
- To reduce the indoor air impacts on asthma, one million children with asthma will have reduced exposure to indoor asthma triggers. In addition, 200,000 low-income adults with asthma, and 2.5 million people with asthma overall, will have reduced exposures to indoor asthma triggers.

NATIONAL ENVIRONMENTAL HEALTH ASSOCIATION (NEHA)—RADON

Working through a cooperative partnership, about 100 NEHA employees are trained annually on indoor air quality (IAQ) and risk reduction strategies. In return, each individual develops a one-year plan of action for achieving IAQ risk reduction as part of their work. The results are impressive.

Tom Dickey, an East Moline, IL local city health department inspector, completed a three-day radon training program in Washington, DC and decided to pursue an incentive-based program for encouraging radon-resistant new construction. He successfully encouraged the City Council to pass a resolution granting a "radon rebate" on the fee that the city assesses on new homes if the homes are built to be radon-resistant. The rebate is roughly proportional to the incremental cost incurred by builders in East Moline to make their homes radon-resistant. Since the rebate program was begun in June 1994, many new homes in East Moline have been built to be radon-resistant. Mr. Dickey has mentored other NEHA/EPA radon community-based risk reduction programs to encourage the use of these kinds of highly-effective incentive strategies.

OAR has recently begun a new initiative on asthma. Its mission is to ensure that indoor environmental management is an integral part of asthma management in the United States. Although both medical treatment and indoor environmental management are needed to effectively control asthma, the latter is not often practiced nor part of the prescription for managing asthma. The Indoor Environments Program will focus on two primary audiences: the public health/medical community and children with asthma and the people who manage their environments. The Program plans to reach these audiences through several activities:

- Health care/managed care summits
- A media campaign
- An in-home education program
- School/day care-based education of children
- Integration of ETS into tobacco control programs
- Improving indoor environments in schools

A number of different specific strategies exist to achieve OAR's priorities. OAR works with its regional offices, state and local agencies, and private partners to get local action on indoor environmental issues. OAR stimulates local action on radon through the State Indoor Radon Grants program, which has resulted in significant risk reduction in homes. A unique feature of the OAR program's voluntary efforts is a network of cooperative partnerships with organizations that speak to and for the public, as well as key constituencies, including county and local environmental health officials, susceptible minority and disadvantaged populations, schools, real estate and building professionals, etc. This network allows OAR to leverage the personnel, expertise, and credibility of these organizations, as well as mobilize hundreds of community-based affiliates at the state and local level.

OAR also takes a proactive approach in providing a broad range of information about indoor air-related risks, as well as the steps to reduce them, through the use of public awareness campaigns, guidance document dissemination, training course delivery, the operation of several linked hotlines and clearinghouses, a web site, and related outreach efforts. These efforts reach a broad audience, including homebuilders and buyers, real estate professionals, health professionals, environmental and public health officials, facility owners and managers, school administrators and teachers, and service providers (such as day care providers, maintenance personnel, and pest control companies).

OFFICE OF PREVENTION, PESTICIDES, AND TOXIC SUBSTANCES (OPPTS)

Many of OPPTS's priorities for 2005 relate to human health indoors. By 2005, OPPTS expects that:

- Lead poisoning will be significantly reduced from levels in the early 1990s, with particular emphasis on children in high-risk groups.
- Of the approximately 3,000 high-volume chemicals in commerce and the 1,000 chemicals expected to enter commerce each year, EPA will significantly increase the introduction and use by industry of safer or "greener" chemicals. Fewer than 100 cases per year will need regulatory management by EPA.
- There will be a significant increase in industry's use of pollution prevention and "green approaches" in the design, development, manufacture, and use of chemicals so that there is increased availability of safer substitutes.
- EPA will annually review about 2,500 Premanufacture Notifications submitted by chemical manufacturers and take appropriate risk management actions to protect human health and the environment. EPA will concentrate on protecting children and workers from potential inhalation and dermal exposures.
- EPA has proposed to amend the TSCA Inventory Update Rule (IUR) to collect information needed for risk screening and develop and implement a chemical hazard classification scheme.
- EPA will achieve significant progress in acquiring test data on chemicals entering commerce and high-volume chemicals, including testing for endocrine disruption.
- There will be significant reductions in exposures to toxic fibers, e.g., asbestos.
- Toxicity test data gaps will be identified for household chemicals which result in substantial exposures to consumers and children. Toxicity testing actions will be initiated or completed for 50 percent of these chemicals. Risk management actions will result in significant risk reduction to consumers, and information/education programs will empower them.

- EPA will improve the ability of the public to reduce exposure to specific environmental and public health risks by making current, accurate, substance-specific information widely and easily accessible.
- EPA will provide chemical data and tools to the public for them to understand and analyze environmental data. The data and tools will be tailored to suit various needs, such as ranking potential concerns for indoor environmental quality and "green design," as well as product labels to be easily understood by consumers.
- All pesticides licensed before 1988 will have complete and reviewed databases, in accordance with the most current requirements, to support their uses (more recently licensed pesticides will already be in full compliance).
- Where necessary, consumer information on labels will be updated and clarified to prevent unnecessary indoor use and exposures.
- For nearly all pesticides, risk assessments accounting for all sources of exposure, including indoor exposures, will be conducted.

OPPTS programs are primarily oriented towards prevention rather than remediation. Both the toxics and pesticide programs operate in an environment of mandated deadlines and regulatory requirements. Science and risk assessment are integral; harmonization of test methods between toxics and pesticides, as well as with others both inside and outside EPA, is an important operating principle. In addition, tool and data development in the areas of exposure, hazard, risk, and economics are ongoing activities in both the toxics and pesticides offices.

OPPTS has regulatory programs in place for two critical indoor pollutants, lead and asbestos. Activities to address these pollutants include:

- Training and certification programs for workers
- State programs and grants
- Information disclosure upon real estate transfer and renovation
- Federally identified hazard levels

CONSUMER LABELING INITIATIVE

The Consumer Labeling Initiative (CLI) is a voluntary, cooperative partnership to foster pollution prevention, empower consumer choice, and improve understanding by presenting clear, consistent, and useful information on household consumer product labels. Government, industry, and other groups are working together in the CLI to make it easier for consumers to find, read, and understand label information about a product's safe use and its environmental and health impacts. This information will enable customers to compare products and safely use the ones they select.

Between 1996 and 1998, CLI conducted significant research with consumers around the country which included one-on-one interviews, focus groups, and phone and written surveys. The purpose of the research was to determine how consumers use pesticide and cleaner labels, if and when they read the labels, and what information they thought could be improved or deleted.

The outcome of the research was a series of recommendations for label improvements in both language and format, as well as the implementation of a consumer education campaign called "Read the Label *First*!" The education campaign is now in full swing and includes the distribution of posters; brochures on the importance of reading labels; promotional items with relevant hotline phone numbers, as well as the campaign logo; fact sheets; TV and radio segments; a truck advertising campaign; and various train-the-trainer sessions given by our state and industry partners, as well as many other CLI participants.

In addition to its regulatory programs, OPPTS also has voluntary pollution prevention activities designed to produce safer indoor environments. OPPTS works with industry stakeholders to develop tools and information that can lead to formulation of safer consumer products for use in the indoor environment. Following Executive Order 13101, OPPTS works with federal consumers, such as the General Services Administration (GSA), to provide them with the information they need to make purchasing decisions that are better for the environment, both indoors and out.

OPPTS also has the Pesticide Environmental Stewardship Program (PESP). PESP is EPA's program designed to address the risks of pesticides and encourage the use of safer pesticides. A major element of PESP is the encouragement of voluntary partnerships with private industry to promote safer pesticides and environmental stewardship.

Consumer education is important and OPPTS is working with partners to clarify product labeling procedures. OPPTS is also working with partners to develop tools to improve the assessment of chemical safety in consumer products and building materials.

OFFICE OF RESEARCH AND DEVELOPMENT (ORD)

ORD produces technical reports, methods, models, and other scientific information to improve the Agency's understanding of the effects of indoor contaminants and their sources, as well as risk management options to reduce exposure. In addition, this research provides technical information that is used by OAR and OPPTS to develop guidance documents on indoor environmental quality and understand the relative risks of various indoor contaminants. The data produced by ORD are also used by product manufacturers to evaluate the risks posed by their products and by building owners and operators responsible for protecting tenants from harmful levels of indoor contaminants. Specific activities planned by ORD are to:

- Develop information on the effects of both biological and chemical contaminants found indoors.
- Develop methods and models to quantify source emissions.
- Collect data on human exposures to indoor contaminants through field studies.
- Produce multi-pathway exposure models that include modules that account for the contribution of contaminants from various indoor microenvironments and take into account penetration of ambient air indoors.
- Develop information to aid school and building managers, the private sector, and government officials in determining which control approaches (e.g., air cleaners, source management, ventilation system design/operation) will have the greatest impact on risk reduction.
- Develop information for manufacturers of building materials and products that pose the greatest risk, assisting them in preventing and reducing emissions through product redesign and process changes.

ORD, OAR, and OPPTS have worked jointly to identify the most critical uncertainties associated with indoor pollutants and have developed the following list of key research needs:

- Source Characterization/Solutions
  - Develop information on and prioritize indoor environmental sources, and establish processes to reduce or prevent pollutant exposures associated with those sources. The most important needs are: (1) prioritization of indoor environmental pollution sources in terms of next actions, e.g., additional studies, guidance development, industry dialogues, and pollution prevention; (2) development of standardized methods for source emission testing; and (3) understanding of typical and high-end indoor exposures, how these exposures relate to indoor pollutant levels, and how their relative risks compare to outdoor air problems and environmental hazards in other media.
  - Assess the impact of building practices on indoor environmental quality. Develop and compare investigation and mitigation techniques, including IEQ and energy performance of ventilation systems in large buildings, cost/benefit analysis of IEQ controls, and assessment of IEQ guidance utilization.
- Health Effects Assessment
  - Improve the Agency's understanding of the health effects of indoor pollutants, both chemical and biological, by developing data on the risks of indoor pollutants, including irritancy, central nervous system and sensory effects, and the effects of mixtures.

- Exposure Assessment
    - Improve the Agency's understanding of the exposure-time-activity pattern factors that contribute to multi-pathway indoor human exposures. Characterize and provide an integrated assessment of these exposures (e.g., inhalation, dietary, dermal) to indoor contaminants and to the dose within the human body, culminating in a first-generation exposure model.
- Risk Assessment
    - Improve the Agency's current knowledge of indoor environmental risks by assessing risks from exposure to chemical pollutants, including organics, nitrogen oxides, carbon monoxide, particulate matter, and microbiologicals.

Current ORD indoor environmental research is conducted as a component of several broader research programs, including particulate matter, air toxics, and human health protection, where indoor exposures contribute significantly to the risks. Research will be conducted using ORD's grants program under the National Center for Environmental Research and ORD laboratories and centers using in-house research facilities and staff, as follows:

Major Research Area	**Responsible National Laboratory**
Source Characterization/Solutions	National Risk Management Research Laboratory (NRMRL)
Exposure Assessment	National Exposure Research Laboratory (NERL)
Health Effects Assessment	National Health and Environmental Effects Research Laboratory (NHEERL)
Risk Assessment	National Center for Environmental Assessment (NCEA)

CONVERSION VARNISH EMISSIONS

ORD completed a study which examines the emissions of formaldehyde and other organic chemicals from conversion varnishes into the indoor environment. Conversion varnishes provide sturdy, chemical- and water-resistant coatings for kitchen and bathroom cabinets and some furniture. They are made up of two components, a polymer resin and a catalyst, which are mixed prior to application. The mixture then reacts to form a continuous film coating on the surface of the wood. The study showed that the organic solvent portion of the coating is emitted quickly, typical of most coatings. These emissions will occur mostly while the cabinet is still in the manufacturing plant. The formaldehyde, however, is emitted by a different mechanism. Rather than showing the emission behavior typical of most coatings, the formaldehyde is emitted over a longer period of time. For the coatings tested, the total amount of formaldehyde emitted was between two and eight times the amount present in the formulation. This reflects a net production of formaldehyde resulting from the chemical reactions that occur during curing and ageing of the coating. In addition, the formaldehyde emissions do not decay as quickly as other (evaporative) emissions more typical of coatings. Rather, the emissions level out over time. The coatings continue to emit significant amounts of formaldehyde even after 42 days, long after they could be placed in a consumer's home. Modeling showed the potential for exposures near the irritation threshold for formaldehyde from this source alone. The next phase of work on this project is to test promising new alternatives to conversion varnishes to determine whether they can reduce total emissions and indoor emissions (and therefore potential for human exposure).

OFFICE OF SOLID WASTE AND EMERGENCY RESPONSE (OSWER)

OSWER's priorities applicable to protecting indoor environments include:

- Improve indoor workplace safety by reducing the risk of industrial chemical accidents. OSWER will develop and disseminate alerts and advisories to industrial sectors based on an enhanced knowledge acquired from increased EPA chemical accident investigations. A joint EPA-OSHA Chemical Accident Investigation Team is currently in place to investigate major chemical accidents and disseminate "lessons learned" to involved industry sectors.
- Reduce risk of worker exposure by reducing the most persistent, bioaccumulative, and toxic (PBT) chemicals in industrial waste streams found at work. By 2005, reduce these types of chemicals in waste streams to 50 percent of 1991 levels.
- Continue to develop and employ innovative strategies for promoting indoor cleanup of contaminants by reducing the cost of waste management without sacrificing human health or environmental protectiveness.
- Continue to provide technical expertise and conduct response actions using Comprehensive Environmental Restoration Compensation and Liability Act (CERCLA) authority. CERCLA authority may be used to respond to threats of environmental releases of hazardous substances, pollutants, or contaminants that are found within homes and offices.
- Expand OSWER's ongoing partnerships with the construction and remodeling industries to promote the use of safe and recycled materials indoors.

PROMOTING LOCAL GREEN BUILDER PROGRAMS

OSWER, in partnership with OPEI and the National Association of Home Builders Research Center (NAHB-RC), is developing a model "green builder" program, based on existing programs in such cities as Austin, TX and Denver, CO. This program will educate builders on environmentally friendly construction and offer them marketing incentives for applying these techniques. The model, which will be designed to be easily adopted by local home builders associations and governments nationwide, will be tested by the Greater Atlanta Home Builders Association as it develops its own "green builder" program. OSWER, OPEI, and NAHB also jointly sponsored the first Green Buildings conference aimed at mainstream home-builders on April 8-9, 1999 in Denver, CO.

Many of OSWER's principles and strategies are designed to reduce risk to humans in the workplace through concepts such as source reduction. OSWER also seeks to reduce future risk inside plants by making unsafe processes safe in the future. OSWER strives to employ good science and technology to make sound environmental policy decisions which are protective and based on common sense and reality. The Office works closely with industry and other stakeholders to assist them in reducing risks to their workers, to their communities, and to the environment by developing pollution prevention and waste minimization tools and ideas. OSWER works to ensure that a high level of public participation is achieved and that state and local involvement exists so that policies and regulations are protective, equitable, and implementable.

OSWER also develops new technologies through research and promotes innovative remediation concepts (such as Brownfields) to achieve the timely, cost-effective cleanup of previously contaminated sites and to develop policy and regulation to prevent future ones from occurring. These cleanup actions seek to minimize threats from exposure to contamination sources whose routes can affect indoor environments (e.g., tap water or indoor air). These pollution prevention strategies, risk management activities, remediation strategies, and chemical emergency response/process safety work, aimed at cost-effectively eliminating, reducing, or minimizing emissions and contamination, will result in cleaner and safer environments in which Americans can reside, work, and enjoy life indoors as well as out.

OSWER also seeks to increase resource efficiency and improve waste management in the construction and demolition industries through the promotion of environmentally friendly building or "green building" programs. While OSWER's primary interest in this field involves expanding recycling and reuse of building products, as well as reducing the amount of demolition debris, the "green building" movement also includes such elements as energy efficiency, water conservation, and indoor environmental quality. Therefore, opportunities exist for OSWER, OAR, and other offices to join forces to create effective, unified "green building" programs that command the respect and interest of the building industries and the public. Through such programs, EPA can further the construction of buildings that protect human health and environmental quality.

OFFICE OF ENFORCEMENT AND COMPLIANCE ASSURANCE (OECA)

OECA's priorities applicable to indoor environments issues are to:

- Work with media program offices to identify areas to be targeted (e.g., high-risk, disproportionately exposed populations and other priority areas of non-compliance).
- Provide the public, especially disproportionately exposed and under-represented populations, with a meaningful opportunity to participate in the development and implementation of environmental protection strategies that involve the National Enforcement and Compliance Program.
- Ensure that all federal and state enforcement programs include a plan for encouraging and responding to citizen reports of violations or other environmental incidents.
- Develop the tools to identify or target particular areas or populations associated with disproportionate exposure and other appropriate factors.
- Work with the Interagency Working Group on Environmental Justice to address case and policy issues that develop between agencies.
- Develop joint agency enforcement initiatives (e.g., EPA/OSHA joint chemical pollution/worker safety cases).

OECA's programs are primarily for the enforcement and implementation of regulatory requirements. In the indoor environments area, OECA is currently focusing on compliance with asbestos in schools requirements, lead-based paint disclosure requirements, and illegal use of pesticides in homes.

However, through the Office of Environmental Justice's (OEJ) Environmental Justice Small Grants and Community/University Grants Programs, OECA has funded numerous local projects dealing with indoor environmental issues, such as lead dust, radon, and asthma. In addition, OEJ has worked closely with OAR to support the Open Airways program and to jointly sponsor a training session on asthma issues and solutions.

OFFICE OF WATER (OW)

Under the Safe Drinking Water Act, EPA sets and enforces standards on public water supplies to prevent human health impact. Human exposures to contaminants brought into the home by drinking water can result from inhalation and dermal exposure, as well as by ingestion via eating and drinking. EPA attempts to take all of these exposure routes into account in the risk assessments that are done for regulatory development.

Inhalation exposure is the major exposure pathway for the risk posed by radon in drinking water. It is also a very significant exposure pathway for other volatile contaminants found in drinking water, such as chlorinated solvents. Inhalation exposure results from showering, in which a large amount of water is aerated in a small enclosed space, as well as from other indoor water uses.

OFFICE OF CHILDREN'S HEALTH PROTECTION (OCHP)

EPA's Office of Children's Health Protection was established in 1997 to promote children's environmental health within EPA, across the federal government, in the non-governmental sector, and in states and communities.

Mission and Goal

OCHP's mission is to make the protection of children's health a fundamental goal of public health and environmental protection in the United States. The Office's goal is that every individual, community, organization, corporation, and government agency will:

1. understand the link between children's health and the environment, and
2. take positive action to improve children's environmental health.

OCHP's overall strategy for addressing risks to children is twofold: (1) to build the infrastructure and capacity to address children's health issues at the federal, state, and community levels and among private sector organizations and individuals and (2) to increase awareness and action on children's environmental health issues throughout all sectors of society.

EPA Science and Regulations

Within EPA, OCHP serves as a focal point for providing technical support on children's environmental health issues to policy makers and outside organizations. It promotes consideration of children's health by media program and research offices, and coordinates Agency-wide initiatives and interagency initiatives with other federal agencies.

In the area of science, OCHP works to improve the science to increase our understanding of children's unique risk and provide sound data on which to base decisions by advocating for increased funding for children's environmental health issues, improved risk assessment procedures, and the conduct of a comprehensive longitudinal cohort study of the relationship between children's health and their environment.

OCHP works to improve the standard-setting process so that risks to children are explicitly considered by providing guidance and analysis on the costs and benefits of protecting children. OCHP works with the National Center for Environmental Economics on indicators of environmental factors affecting children's health.

Federal Leadership

EPA has exercised a leadership role in the federal community on children's environmental health, in part by recognizing the fundamental importance of a top to bottom partnership with the Department of Health and Human Services and other agencies through the President's Task Force on Environmental Health Risks and Safety Risks to Children.

States

OCHP provides resources and assistance to the states to develop programs to address their children's environmental health issues through state organizations such as: the Association of State and Territorial Health Officials (ASTHO); the Environmental Council of the States (ECOS); the National Conference of State Legislatures (NCSL); and the National Governors Association (NGA). In addition, OCHP provides resources to EPA's Regional offices to support their efforts to build capacity in the states and local communities.

Communities

OCHP works with community organizations to help them understand and address their children's environmental health issues. Examples include: (1) the Child Health Champion Community Program to empower local citizens and communities to take steps toward protecting their children from environmental health threats; (2) the Child Health Champion Environmental Monitoring and Education Project to provide easily understood up-to-date environmental information for communities; and (3) working with youth organizations, such as the Boy Scouts and Girl Scouts, 4-H, the Future Farmers of America, and the United National Indian Youth, to incorporate children's environmental health into their existing programs. OCHP maintains EPA's Children's Health Protection Web Site, which provides information to parents and others on ways to protect children from environmental risks.

Private Organizations

OCHP works with private sector organizations on children's environmental health issues. For example, OCHP is working with health care provider organizations, such as the American Academy of Pediatrics and the American Nurses Association, to promote the incorporation of environmental health into pediatric and nursing practices to increase the ability of primary health care providers to identify, prevent, and reduce environmental health threats to children.

OFFICE OF ADMINISTRATION AND RESOURCES MANAGEMENT (OARM)

The Office of Administration and Resources Management's main goal regarding human health indoors is to provide a safe and healthful environment for EPA's own employees. Because of their expertise within EPA, OARM often works in conjunction with other EPA offices, other federal agencies, and outside organizations on indoor environmental issues.

- As part of the EPA New Headquarters project, OARM performs chamber testing, modeling, and specification writing to strive for improved indoor environmental quality. The protocols that have been developed by OARM for office furniture during this process are now being used in an Environmental Technology Verification (ETV) project with the furniture industry that will result in a national furniture testing program.
- OARM is also actively working with OPPTS to institutionalize the "Green Cleanser" project and develop language to promote the use of these cleaners in EPA buildings.
- With Public Technology, Inc. (PTI), OARM participated in publishing two guides for sustainability in buildings. OARM is continuing to work with the President's Council on Sustainable Development on these and related projects.
- With the General Services Administration, OARM has developed guides for the management of asbestos and lead at federal facilities.
- OARM's multimedia laboratory uses computer technology to build learning and program support tools that have wide application in the federal, private, and academic communities.

REGIONAL OFFICES

The EPA Regions support and implement the national programs discussed earlier in this Appendix. In doing so, these offices have demonstrated initiative and creativity in working with very limited resources to address indoor risks in innovative ways.

When available, the Regions use statutory authorities. For example, EPA Regional Offices are:

- Working with state and tribal partners to develop lead programs, per Title IV of the Toxic Substances Control Act, for certification and training of lead workers.
- Working with state and tribal partners to implement radon programs using the grant authorities of the Indoor Radon Abatement Act to promote voluntary programs for radon awareness, testing, and mitigation.
- Working with state and tribal partners to develop and implement asbestos-in-schools management programs, per the Asbestos Hazard Emergency Response Act.
- Working with public water supplies to address the requirements of the Safe Drinking Water Act and the Lead and Copper Rule.

Of equal, and in some cases more, importance are unique efforts for outreach, education, and technical assistance for non-regulatory programs, using a multitude of government, non-profit, and other stakeholder partners. Examples include:

- Through Regional Indoor Air Quality Programs, efforts have been tailored to educate the public on a variety of issues using an increasing array of effective tools. Depending on geography and climate, such issues as toxic mold, asthma and its triggers, CO poisoning, indoor use of pesticides, and environmental tobacco smoke are being addressed.

REGION 1
NH TOOLS FOR SCHOOLS

Two schools in New Hampshire were the first successful pilot schools in the country to fully and successfully implement EPA's Tools for Schools Indoor Air Action Kit. The Pennichuck Junior High School in Nashua and the Little Harbor Elementary School in Portsmouth began the process of implementing the Kit in the Fall of 1996 by appointing an IAQ Coordinator and forming an IAQ Team. To better inform team members about indoor air quality and how to more fully use the guidance, each team was given indoor air training by the NH Division of Human Health Services and an EPA grantee, the NH Coalition for Occupational Safety and Health (NH COSH).

REGION 2
CLINICAL DIRECTORS NETWORK PROJECT

It is important to research effective ways to reduce asthma morbidity and translate that research into practice. Translation is the focus of a Region 2-funded grant to a non-profit entity called Clinical Directors Network (CDN). CDN's asthma project is operating in 11 sites in four EPA Regions nationwide. It is a clinically controlled study designed to measure the improvement in asthma morbidity that can be gained from implementing both in-home environmental interventions and improved clinical management of asthma. It makes use of the "Best Practices" known so far, and will offer insight as to the best ways to implement this information.

- IAQ in Schools is being addressed through large outreach campaigns using EPA's IAQ Tools for Schools Kit. Leveraging of stakeholder resources is crucial to these efforts.
- Exposure to lead from paint, dust, soil, and drinking water are addressed with large outreach and education campaigns, incorporating many partners.

REGION 10
SEATTLE/ALASKA HEALTHY HOMES

Through a grant with the American Lung Association of Washington, the Master Home Environmentalist Program promotes human health by increasing awareness of home environmental pollutants and encourages actions to reduce exposures. The MHE program uses innovative and holistic approaches to identify hazards and ways to make homes healthier. The program relies on volunteers to reach out to local communities to deliver the latest information about environmental health issues. Volunteers complete extensive training in lead, dust, indoor air, household hazardous chemicals, and moisture problems in the home. Outreach has been conducted in Galena, AK and Seattle, WA. There is a new program beginning in Yakima, WA.

Appendix **B**

PARTNERS IN INDOOR ENVIRONMENTAL PROTECTION

OTHER FEDERAL AGENCIES

WORKING TOGETHER

EPA has worked in conjunction with a number of federal agencies on joint efforts to protect human health indoors. A few examples of these activities are:

JOINT PUBLICATIONS:

With CPSC:

- The Inside Story
- Asbestos in the Home
- Combustion Appliances and Indoor Air Pollution
- What You Should Know About Using Paint Strippers
- Indoor Air Pollution: An Introduction for Health Professionals

With DHHS:

- Building Air Quality: A Guide for Building Owners and Managers
- A Citizen's Guide to Radon
- Introduction to Indoor Air Quality

With CPSC and HUD:

- Protect Your Family From Lead in Your Home

With CPSC, OSHA, the National Institute for Occupational Safety and Health (NIOSH), and the Colorado Department of Public Health and the Environment (CDPHE):

- Preventing Carbon Monoxide Poisoning from Small Gasoline-Powered Engines and Tools

JOINT PROJECTS:

EPA is working with:

- HUD, CPSC, and DHHS on a number of projects related to lead-based paint hazards.
- GSA, under Executive Order 13101, to develop guidance on environmentally preferable products for use in federal buildings.
- The Department of Energy (DOE) to implement energy-efficiency and other improvements at EPA facilities to improve laboratory operations and to take relevant lessons to a broader audience (e.g., hospitals, computers, etc.).
- The Department of Agriculture (USDA) to develop a list of bio-based products that may be preferable for use in the indoor environment.

A number of federal agencies and departments outside EPA have responsibilities for protecting human health indoors. The efforts of the Occupational Safety and Health Administration in the Department of Labor (DOL), the National Institute for Occupational Safety and Health in the Department of Health and Human Services, and the General Services Administration focus on protecting the health of the workforce.

Other agencies and departments focus on reducing exposures and risks to the general population. The Consumer Product Safety Commission is responsible for protecting American families, especially children, from the unreasonable risk of injury (including illness) and death from about 15,000 types of consumer products under the Commission's jurisdiction. The Department of Housing and Urban Development provides for safe and healthful housing through programs to reduce exposures to formaldehyde, lead, and other toxic materials in homes.

Several agencies and departments are involved in important research activities to ensure the protection of human health indoors. For example, the Centers for Disease Control and Prevention (CDC), the Agency for Toxic Substances and Disease Registry (ATSDR), and the National Institutes of Health (NIH) perform wide-ranging public health research on pollutants indoors (e.g., lead, radon, environmental tobacco smoke, combustion products, allergens, Legionnella). The Department of Energy evaluates the health effects of radon exposure. DOE and the National Institute of Standards and Technology (NIST) perform research on the relationship between air movement and contaminant levels in buildings and other related issues.

A number of other federal agencies and departments also have key roles in protecting human health indoors. Many of these agencies are members of the Interagency Committee on Indoor Air Quality. A list of Current Federal Indoor Air Quality Activities can be found in EPA publication EPA-402-K-99-001.

OCCUPATIONAL SAFETY AND HEALTH ADMINISTRATION (OSHA)

Under the Occupational Safety and Health Act (OSHAct), OSHA develops and enforces occupational safety and health standards, including those related to exposures to toxic substances, and has proposed a comprehensive indoor air quality standard for workplaces.

Key activities at OSHA include:

- Protecting the health and safety of American workers by promulgating mandatory standards and by inspecting workplaces to ensure compliance with those standards.
- Regulating worker exposure to toxic substances and harmful physical agents, including asbestos, lead, and noise.
- Publishing a proposed rule on April 5, 1994 (59 FR 15968) to require employers to write and implement indoor air quality compliance plans that would include inspection and maintenance of current building ventilation systems to ensure that they are functioning as designed. Other proposed provisions would require employers to maintain healthy air quality during renovation, remodeling, and similar activities. The provisions for indoor air quality would apply to 70 million workers and more than 4.5 million non-industrial work environments, including schools and training centers, offices, commercial establishments, health care facilities, cafeterias, and factory break rooms. The OSHA Standards Team is analyzing the docket, defining issues, and gathering new data.
- Assisting and providing guidance to federal and state compliance officials, and to building managers, employers, engineers, and owners through OSHA's outreach program, in evaluating indoor environmental quality in non-industrial workplaces (including the occurrence of Legionnaires' disease and occupational asthma).

Consumer Product Safety Commission (CPSC)

The Commission enforces five federal statutes: the Consumer Product Safety Act, the Flammable Fabrics Act, the Poison Prevention Packaging Act, the Federal Hazardous Substances Act, and the Refrigerator Safety Act. CPSC's mission is to:

- Protect the public against unreasonable risks.
- Assist consumers in evaluating comparative safety.
- Develop uniform safety standards so as to minimize conflicting state and local regulations.
- Promote research and investigation into causes and prevention of product-related deaths, illnesses, and injuries.

CPSC uses a variety of approaches to identify product hazards, including an internationally recognized hospital emergency room reporting system and a toll-free hotline. The Agency assesses these hazards using a risk-based approach grounded in the best scientific data. Once the hazards are assessed, CPSC uses a wide range of tools to correct them, including:

- Voluntary standards and guidelines
- Product recalls and corrective actions
- Mandatory rulemaking (e.g., performance standards, bans, labeling)
- Consumer education

Because CPSC is a federal agency, its product safety work and uniform safety guidance and standards ensure businesses a level playing field for domestic and imported consumer products. CPSC evaluates and acts on health hazards associated with the use of products in the following areas: fire (e.g., cigarette lighters and upholstered furniture); mechanical (e.g., children's products, household/structural products, power tools and equipment, sports and recreational products); electrical (e.g., lights); and chemical (e.g., fuel-burning appliances). Some specific CPSC activities relating to human health indoors include:

- Evaluating carbon monoxide alarms to protect against CO poisoning and working with Underwriters Laboratories (UL) to develop a new standard for CO alarms.
- Conducting recalls and developing corrective actions for products that present a substantial risk of CO poisoning.
- Developing voluntary standards to limit combustion pollutant emissions from kerosene heaters, unvented gas space heaters, and camping heaters.
- Working with the gas water heater industry to develop an effective voluntary standard to address the ignition of flammable vapors.
- Evaluating consumer products for the presence of asbestos fibers and assessing any potential risk.
- Reducing consumer exposures to lead and protecting against childhood lead poisoning by investigating the release of lead from imported vinyl miniblinds and requesting the industry cease using lead as a stabilizer in these products.
- Conducting recalls of children's jewelry and toys containing lead.
- Initiating rulemaking to ban candle wicks containing greater than 0.06 percent lead.
- Conducting a study of leaded paint and developing a strategy for use by state agencies for identifying and controlling leaded paint.
- Assessing the potential toxicity of fire-retardant chemicals.
- Initiating rulemaking to address the hazards of small flame ignitions of upholstered furniture.

- Assessing the performance of residential smoke alarms.
- Identifying the potential for emissions of bioaerosols from portable humidifiers and developing guidelines for cleaning and maintaining these humidifiers to reduce bioaerosol emissions.
- Investigating and analyzing monitoring data on biological pollutants in homes, as part of the Harvard Six-City Study.
- Assessing the impact of selected residential heating, ventilating, and air-conditioning systems and control technologies on indoor air quality.
- Evaluating carpet systems to determine if the chemicals they emit into the air might be responsible for the health complaints reported by consumers.
- Measuring and assessing the risk of indoor air pollutant emissions from wood-burning stoves.
- Convening a Chronic Hazard Advisory Panel of scientists to study issues related to the chronic toxicity and risk associated with exposure to diisononyl phthalate (DINP) in children's PVC products.
- Promulgating several regulations requiring child-resistant packaging for medicines and household chemicals to reduce the number of deaths to children under the age of five from accidental ingestion.
- Assessing the potential for noise-induced hearing loss from consumer products.
- Developing and disseminating consumer information booklets/ brochures on asbestos, formaldehyde, biological pollutants, lead, combustion pollutants, and carbon monoxide alarms.

DEPARTMENT OF ENERGY (DOE)

Under the Energy Organization Act, the Atomic Energy Act, and the Energy Conservation and Production Act, DOE is charged with:

- Conducting research on the health effects of ionizing radiation, including radon.
- Establishing guidance for energy-efficient buildings and promoting their use.
- Evaluating the impact of energy conservation standards on habitability.

Key research at DOE includes:

- Developing, testing, and evaluating energy-efficient and cost-effective techniques to maintain indoor environmental quality.
- Developing methods and protocols for measuring emissions from key building materials and products.
- Determining the relationship between organic pollutants in large buildings and residences and energy-conservation methods.
- Developing methods to model and measure infiltration and interzonal airflows and assess ventilation of U.S. housing and associated energy use.
- Assessing the potential to improve productivity of office workers by providing better indoor environments (in conjunction with the National Institute for Occupational Safety and Health).
- Supporting the American Society of Heating, Refrigerating, and Air-Conditioning Engineers (ASHRAE) in developing effective ventilation and indoor air quality standards.
- Developing practical measurement techniques for ventilation rates and efficiencies.

Department of Health and Human Services (DHHS)

Under the Public Health Services Act (PHSA), DHHS performs research and other activities on the cause, diagnosis, treatment, control, and prevention of disease related to indoor pollution. These activities include:

- Identifying pollutants and other environmental conditions responsible for human disease and adverse effects on humans.
- Evaluating the health costs of pollutants (with EPA and others).

There are a number of institutes and agencies within DHHS that are doing work to protect human health indoors. The National Institute for Occupational Safety and Health within the Centers for Disease Control and Prevention answers inquiries on indoor environmental quality in non-industrial workplaces (e.g., offices) and performs site investigations to solve environmental problems in these workplaces. NIOSH also conducts epidemiologic research on the causes and prevention of health effects in non-industrial indoor workplaces and, through the Indoor Environment Team of the National Occupational Research Agenda process, is working to define and facilitate a priority research agenda to improve the health of workers in these indoor environments.

Other activities at CDC include:

- Providing information to state health departments and members of the public concerning the health effects of indoor environmental pollutants.
- Developing reliable tests for tobacco smoke exposure.
- Providing assistance to state and local health departments in conducting screening and surveillance activities in order to minimize the adverse effects of environmental lead contamination.
- Providing assistance to states to address asthma including a variety of training and program development efforts, transfer of best practices, and modes for surveillance.

The Agency for Toxic Substances and Disease Registry within DHHS provides assistance and advice on indoor environmental contaminant exposures related to hazardous waste sites.

Several institutes within the National Institutes of Health are also doing work to protect human health indoors, especially in the area of asthma.

DEPARTMENT OF HOUSING AND URBAN DEVELOPMENT (HUD)

HUD has been actively involved in a number of key indoor environmental issues. Through the National Manufactured Housing Construction and Safety Standards Act, HUD has provided for safe and healthful conditions in manufactured housing. Standards for formaldehyde emissions from pressed wood products have been promulgated under this Act. In addition, HUD is working to improve the air distribution systems in these types of homes. HUD also determines HUD/FHA and Public and Indian housing policies on radon issues.

Through the Residential Lead-Based Paint Hazard Reduction Act, HUD works to reduce lead exposures in U.S. housing. Key activities in this area have included:

- Implementing the HUD Lead Hazard Control Grant Program, which has eliminated lead-based paint hazards in 50,000 privately owned low-income housing units.
- Scientifically evaluating the effectiveness of a range of lead hazard control strategies, through the National Evaluation of the HUD Lead Hazard Control Grant Program.
- Conducting research on lead-based paint identification, evaluation, and control methods, and conducting the National Survey of Lead and Allergens in Housing, the first such survey.
- Developing, implementing, and enforcing (with EPA and DOJ) the Lead Disclosure Rule for renting or selling pre-1978 housing.
- Developing, implementing, and enforcing the Lead Safe Housing Rule for federally owned and assisted properties.
- Surveying public awareness of lead-based paint hazards and their causes, and determining the effects of hazard disclosure on real estate transactions.
- Developing model provisions for state and local housing codes, model

abatement specifications, and work practice guidance and training curricula for reducing lead-based paint hazards during maintenance, renovation, remodeling, and rehabilitation work.

Finally, conducting general or targeted community programs on environmental health and safety hazards under their "Healthy Homes Initiative," HUD considers allergens and asthma, asbestos, combustion products, insect and rodent pests, mold and moisture, pesticide residues, and radon key targets for intervention.

A key Initiative objective is reducing multiple hazards in housing that pose risks for residents, particularly children, using a single intervention. Healthy Homes projects also focus on developing and implementing cost-effective strategies for hazard assessment and intervention methods, and for developing and disseminating technical assistance, guidelines, and model provisions for housing codes and standards. Much of this work is accomplished through competitive grants to communities to support local Healthy Homes programs. HUD also works with other federal agencies to fund joint Healthy Homes research and education projects, such as "Help Yourself to a Healthy Home," a booklet that contains tips on improving indoor air quality.

General Services Administration (GSA)

GSA provides indoor environmental quality guidelines for federal (GSA-owned) buildings and leased space. Key activities include:

- Providing information to consumers who purchase office furniture from the Federal Supply Service schedule.
- Coordinating radon testing and mitigation in GSA-controlled buildings.
- Funding indoor air quality research by the National Institute for Occupational Safety and Health in GSA buildings.
- Providing an indoor environmental quality program that has:
  - An ongoing component that includes responding promptly to concerns and, when possible, correcting problems discovered.
  - A pro-active component that includes conducting indoor environmental assessments as part of a survey program.

The Federal Consumer Information Center (FCIC), an office of GSA, offers several publications from federal agencies on indoor environmental quality in their Consumer Information Catalog and on their web site at http://www.pueblo.gsa.gov.

NATIONAL INSTITUTE OF STANDARDS AND TECHNOLOGY (NIST)

The focus of studies at NIST has been on the relationship between ventilation and contaminant levels in buildings. Activities at NIST include:

- Developing models to account for air movement and contaminant dispersal in buildings.
- Developing an understanding of factors affecting the mixing of carbon monoxide in buildings as it relates to the location of alarms.
- Performing simulation studies of approaches to the ventilation of residential buildings.
- Developing test methods and procedures for studying air change characteristics, pollutant levels, and their relationship in large buildings.
- Developing a practical guide to procedures for assessing ventilation rates in commercial buildings.
- Developing test methods for lead in paint (in cooperation with HUD and EPA).
- Maintaining national radium and radon measurement standards.

STATE, LOCAL, AND TRIBAL AGENCIES

The quality and structure of state, local, and tribal indoor environmental programs vary significantly from state to state. Some states have strong or moderately strong indoor environmental programs, while others have essentially no programs at all.

State, local, and tribal programs provide public information, problem assessment, and response, but often these activities are divided among several agencies, particularly at the state level, as a reflection of the multifaceted nature of indoor environmental issues. Some states, like California, Florida, and Vermont, have interagency indoor environmental groups to coordinate activities across state agencies. The strongest state programs are those which have been mandated by state legislation. In these states, much of what has been achieved has been through voluntary compliance. Many state and local governments do have some regulatory authority in specific areas (e.g., asbestos, lead, radon, environmental tobacco smoke). A substantial component of many state programs is to assist local governments and tribes to address indoor environmental issues at the local and tribal level.

State-level indoor environmental programs are often hampered by the lack of a routine funding mechanism, with the exception of state radon programs which can receive federal funding. Agencies sometimes respond to problems identified through publicity or public outcry. Such response is frequently reactive and crisis-driven. In some states, there is no organized structure in place to educate or empower the public about their indoor environment, and funding may decrease when the issue drops out of the media spotlight.

Like state governments, local health and/or environmental offices often have no established indoor environmental programs. They may create a mechanism to respond to a current crisis, routine public inquiries, or public outcry.

The scarcity of local government programs is being offset by grassroots coalitions and non-profit organizations, extension educators, and local professional organizations working independently and in cooperation with each other and federal, state, and local officials on public outreach and program implementation. Some funding and training for indoor environmental activities is available for and utilized by local governments, local health and environment officials, and non-profit organizations. More limited funding may be available for local indoor environmental needs on a competitive or ad hoc basis from federal and/or state agencies.

Some tribal governments have established radon and indoor environmental programs and receive federal funding. The close-knit nature of tribal councils and the high regard of elders have proven effective in implementing grassroots environmental programs and allowed for good coordination of environmental program activities. Economic and cultural issues make some environmental issues a particular challenge. Involvement of the tribal council and elders assures awareness of cultural sensitivities and increases the opportunity for success.

The strongest indoor environmental programs were found in states where there was a funding mechanism, upper management support of the program, and/or full time staff dedicated to indoor environmental efforts. However, even states with strong programs generally face constraints which keep them from doing the kind of proactive outreach which would prevent indoor environmental problems and crises or have serious gaps in their programs. For example, some statewide/regional programs cover such a large geographic area that individual city or county assistance could be more effective. Some states have a strong indoor environmental quality program in schools, but do not address homes at all. In other states, efforts for lead or pesticides may be targeted to specific audiences due to staff limitations (e.g., integrated pest management in schools, lead awareness to real estate professionals). Pesticides programs are frequently housed in state agriculture departments, which will follow up on indoor environmental concerns regarding pesticides misuse if contacted.

OTHER STAKEHOLDERS

Many different entities in the private sector impact the state of human health indoors. A few of the key stakeholder groups that have a role in protecting human health indoors, and their potential roles in solving indoor environmental problems, are discussed below.

Consumer, Environmental, and Health Professionals

Consumer, environmental, health, and other professionals are knowledgeable about the symptoms and effects produced by environmental pollutants and can advise the public on possible mitigation of environmental exposures. They use diverse approaches to protecting human health indoors, including developing information and education programs to educate the public about indoor environmental quality, conducting research to identify problems and recommend solutions, and participating in the policy-making process.

Manufacturing and Natural Resource Industries

Manufacturers can ensure good indoor environments by designing products and materials that eliminate or reduce exposures to toxic chemicals, pesticides, and other pollutants to safe levels. These include consumer and commercial products, building materials, office equipment, and furniture. Manufacturers can also label their products so that they will be properly used and maintained. If a supplier provides raw materials (e.g., chemicals) to be formulated further into a product, the supplier can provide the formulator with sufficient health and safety information to allow the formulator to determine if the raw material can be safely used in the intended application. Manufacturers and suppliers can conduct research and adopt test procedures (e.g., emission test procedures) and standards to ensure that the products and materials that they sell are safe for use in indoor environments.

Builders and Architects

Builders and architects can work to design and build structures that eliminate indoor environmental problems or enhance indoor environments. By thinking about the quality of the indoor environment in the design stage, in construction practices, and in remodeling, builders and architects can have a substantial impact on the health and safety of the building occupants. Builders and architects can help achieve safe indoor environments by selecting building materials that will not release harmful levels of toxic chemicals into occupied indoor environments (either when the materials are new or as they age) and by designing buildings to be in compliance with indoor air quality ventilation standards. During the remodeling of occupied buildings, builders and architects can help protect the safety of tenants by isolating them from pollutants generated during construction work.

Building Owners, Managers, and Engineers

Building owners, managers, and engineers ensure good indoor environmental quality by properly operating and maintaining buildings. Building owners, managers, and engineers can foster a good indoor environment by adopting maintenance procedures to eliminate and prevent contamination and ensure an adequate supply of clean air to occupants; using zone ventilation or local exhaust for indoor sources; developing specific procedures for use of cleaning solvents, paints, pesticides, and other products and materials within the building; and abiding by recognized standards of care for building maintenance. Their role includes establishing a process to educate building occupants about their roles in maintaining good indoor environmental quality and encouraging an active exchange of information about indoor environmental problems. They can develop and adopt formal protocols to investigate indoor environmental complaints from occupants, thereby encouraging an atmosphere of trust.

Diagnostic and Mitigation Services

Diagnostic and mitigation firms respond to hazards and complaints in problem buildings. They may work closely with building owners, managers, and engineers or individual homeowners to investigate indoor environmental quality issues. Professionals in these firms span a broad range of occupations, including industrial hygienists, mechanical (ventilation) engineers and technicians, microbiologists, architects, chemists, air pollution scientists, industrial engineers, risk assessment personnel, abatement personnel, and others. The services of most of these firms include evaluations of ventilation systems, measurement of indoor pollutants, and characterization of the sources of pollutants in buildings. Through these efforts, they can be instrumental in turning a problem building into a healthy building.

Real Estate Industry

The real estate industry has begun addressing a variety of indoor environmental issues in the past few years as a result of both client demand and legal requirements. The real estate industry, discovering a need to know more about radon, lead, asbestos, the safe application of pesticides, and underground storage tanks, is partnering with government and industry organizations to provide the necessary training to its members to facilitate transactions and improve customer service.

Unions

Unions can protect human health indoors by ensuring a clean and healthy indoor environment for their members. They can work with building owners, managers, and engineers to ensure that employees are afforded an optimum work environment. If problems occur, they can come to the aid of employees who feel that they have been improperly exposed to pollutants in their workplaces and can work with building designers, owners, managers, and engineers in the design and operation of healthy workplaces.

STANDARD-SETTING ORGANIZATIONS

Standard-setting organizations (e.g., building code organizations, the American Society for Testing and Materials (ASTM), the American National Standards Institute (ANSI), the American Society of Heating, Refrigerating, and Air-Conditioning Engineers, Underwriters Laboratories, NSF International, the American Conference of Governmental Industrial Hygienists (ACGIH)) can play an important role in protecting human health in indoor environments. Depending on the organization, they can provide a range of services. One important service of some standard-setting organizations is to foster healthy indoor environments by developing or enhancing, providing for efficient use of, and, and in some cases, enforcing model building codes. Other services standard-setting organizations may provide are setting uniform methods of testing, establishing levels of accepted practice, or developing and maintaining consensus standards. Some may provide certification opportunities, laboratory testing and toxicological assessments related to certification, and conformity assessments and compliance monitoring. Education and training services may also be provided. Standard-setting organizations can help product manufacturers, code writers, designers, builders, enforcement officials, and others perform their functions in a more effective and efficient manner. Standard-setting organizations can also play an important role in providing the public with some assurance that their homes, schools, and workplaces, and the products that go into them, are safe.

RESEARCH ORGANIZATIONS

Many research organizations work to protect human health indoors. Some of these organizations address policy issues, such as providing critical analyses of the potential risks for pollutants indoors, addressing land use and building design issues, or setting future strategies for protecting indoor environments. Scientific research organizations address a wide range of issues related to indoor environments, including proper building design and operation, health and comfort impacts of poor indoor environments, measurements of indoor pollutants and the characterization of emissions from products and materials used indoors, and exposure mitigation (e.g., ventilation, air cleaning, source control, individual behaviors). Because research in indoor environments is relatively new, these organizations play a key role in determining future areas of concern for indoor environments.

Individuals

Individuals are the strongest force in protecting human health indoors. Consumers protect their own health and the health of those around them by properly maintaining their homes and making informed choices about consumer goods and services. Building occupants (e.g., office workers) do the same by properly using products and equipment within the building. With knowledge, individuals can take many actions to avoid personal exposures. The value of individual behavior cannot be minimized in our efforts to develop and implement a nationwide strategy to improve indoor environments.

Appendix C

COMMENTS ON THE DRAFT REPORT

Access to Comments in the OAR Docket

Full comments on the draft report can be accessed in docket number A-98-05 at:

Air and Radiation Docket and Information Center
Mail Code 6102
Room Number M1500
401 M Street, SW
Washington, DC 20460

Phone: 202.260.7549
Fax: 202.260.4400
E-mail: a-and-r-docket@epa.gov

Full comments on the draft HBHP report can be accessed through the OAR docket (see the box on the right). A summary of the comments is provided below. Comments were received from over 40 individuals and organizations and represented a wide array of perspectives.

A number of state officials acknowledged how important the indoor environment is to public health, and that increased attention and resources are needed at the state and local level in order to effect positive change. We agree with this, and hope that state and local governments, as resources permit, seriously consider implementing some of the potential actions contained in this report. In addition, EPA and others need to partner with state, local, and tribal governments as we begin to take action to improve the indoor environment. Moreover, federal and state legislators need to consider funding sources for state, local, and tribal involvement.

Several commentors wanted a more explicit recognition of the interrelationship between indoor and outdoor air pollution, and the important contribution that ambient pollution makes to the indoor environment through natural and mechanical ventilation. We recognize this important relationship, and have modified the principles outlined in the final report to more fully make this connection. Moreover, the Office of Air and Radiation at EPA has recently begun an air toxics pilot project in the City of Cleveland to take an integrated look at both the outdoor and indoor sources of air toxics, given the strong interrelationship between outdoor and indoor air. When this pilot is successfully completed, EPA hopes to replicate it in other urban areas across the country.

A number of commentors wanted us to more explicitly recognize that combinations of pollutants (i.e., mixtures) may also be responsible for poor indoor environmental quality. We agree with this; the draft report specifically recognizes the dearth of research on mixtures and calls for such research as part of any cross-Agency research strategy in the "Potential Actions" section associated with Goal 1.

Some commentors wanted us to add much more specificity to the "Potential Actions" section. For example, some commentors wanted us to specify particular building types, and, for each type, detail associated actions. We intentionally avoided this type of approach because certain universal needs exist across all building types. For example, under Goal 2, generating good cost/benefit data, creating integrated designs, marketing attractive incentives, and promoting good IEQ standards are common to all building types.

A number of commentors expressed concern that there were not more distinct references in the draft report to the importance of ventilation to good IEQ. We agree with the commentors on the importance of adequate ventilation and good IEQ. While not explicitly addressed in many places in the draft, we believe that ventilation issues will be addressed by several of the potential action areas. For example, ventilation issues will be captured in "risk management research" under Goal 1, and are an integral part of "excellent IEQ standards of care" under Goal 3. However, we have made several changes in the final report to more specifically recognize the important link between ventilation and IEQ.

Finally, some commentors pointed out that a section was needed to address the legislative and regulatory (including building code) changes that will be necessary to realize the goals outlined in the draft report. We acknowledge that many approaches, voluntary, as well as regulatory, will be needed in order to achieve healthier indoor environments. Those who embrace the vision and goals of the HBHP report will need to decide the most effective approach to implement the potential actions identified.